REPAIR YOUR OWN

SADDLERY AND HARNESS

A Step-by-Step Guide

REPAIR YOUR OWN SADDLERY AND HARNESS

A Step-by-Step Guide

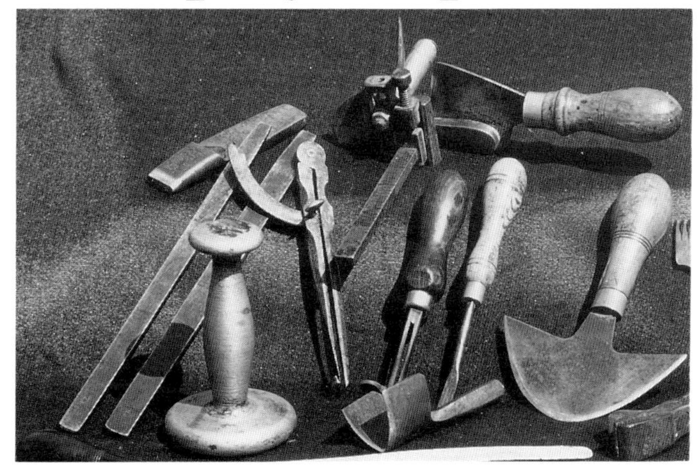

ROBERT H. STEINKE

J. A. ALLEN
LONDON

British Library Cataloguing in Publication Data
A catalogue record for this book is available from the British Library

ISBN 0-85131-597-6

Published in Great Britain in 1994 by
J. A. Allen & Company Limited,
1 Lower Grosvenor Place, Buckingham Palace Road,
London, SW1W 0EL.

Typeset in Hong Kong by Setrite Typesetters Ltd.
Printed in Spain by Printeksa, S.A.

All photography and illustrations by Robert H. Steinke
Designed by Judy Linard

This book is dedicated to

Jenny, my wife, for her belief in this book

and

Duke, my faithful old cob, who has taught me so much over the last twenty years

CONTENTS

PREFACE

In writing this book, I may well have used terms for saddlery or workshop tools and practices that may not be common to all of you; you may well know to what I am referring, but under a different name.

The difference in terminology was brought home to me early on in my career while attending the Cordwainers College in London. My fellow students were a wonderful collection: Jock was a serving soldier; Robert was from a racing background; Dave worked in the City of London; the other Dave worked in the Home Counties; Mark came from the country in Devon; and I was apprenticed to Keith Luxford, a true showman.

Our poor instructor, Norman Evans, had his work cut out with the clash of cultures, and the fact that all the students had different names for even the most simple of tools or jobs.

I have included a glossary to help with any terminology queries, and hope that this book will help make your first steps in saddlery and harness repairs much easier.

Robert Steinke

ACKNOWLEDGEMENTS

I would like to thank these people.

Keith Luxford of Teddington, my first and only employer in the saddlery trade. As well as nurturing my interest in the craft, he also emphasized the need for design, flair and a love of the horse.

Jim Roberts, now retired, a Welshman who taught me at the bench. We never made a larger batch than three saddles at a time in the workshop and this enabled me to see the make-up of a saddle in more detail.

The team at J. A. Allen & Co., especially Caroline Burt for her ideas and encouragement, Jane Lake for editing my manuscript and for her time spent taking me through this book, and Judy Linard for her design work.

And, finally, my wife, Jenny, who encouraged me to write this book which has been in my head for many years. She is a good companion in the workshop and a willing typist.

METRIC CONVERSION TABLE

Most, if not all, manufacturers of buckles, rings, rubber rein grips and other fittings for the saddlery trade still market their products in imperial, or British, sizes, but leather is traditionally measured in imperial for area and metric for thickness. For example, bridle butts might be sold as 5 ft long × 3.5 mm thick and shoulder leather as 3.0 mm thick × £× per square foot. All measurements in this book are, therefore, in imperial sizes except for hide thickness sizes which are metric.

Plough gauges are marked in divisions of ⅛ of an inch, so if you prefer to work in metric sizes you will need to measure the gauge with a separate ruler or re-mark the gauge bar yourself.

The following 'workshop' conversion table (used by at least one major supplier) is not strictly mathematically correct, but it provides a useful guide when selecting fittings.

⅜ in = 10 mm
½ in = 13 mm
⅝ in = 16 mm
¾ in = 18 mm
⅞ in = 23 mm
1 in = 25 mm
1⅛ in = 29 mm
1¼ in = 32 mm

Mathematically 1 in = 25.4 mm (2.54 cm) and 1 ft = 30.48 cm (0.3048 m)

INTRODUCTION

Over the past few years, the cost of good quality saddlery has risen sharply. This is due to, amongst other factors, the rise in oil prices (which affects the dressed hide) wage settlements and property prices. It is, therefore, very important to look after the equipment we use and prolong its life by repairing it when necessary. Unfortunately many saddlers would prefer to sell customers a new item than repair an old one. This is not to say that saddlery does not wear out, but well-made and well-fitting tack can have its useful life extended by sympathetic repairing as well as regular cleaning. By sympathetic repairing, I mean a neat hand-stitched splice or addition, instead of the rivet or nut and bolt repairs we have all seen.

Many of you may think that you could not manage to hand-stitch such thick leather, or that repairs should only be carried out by qualified saddlers and not by the layman. Let us take each of these doubts in turn.

Firstly, I believe nearly anybody who is practical enough to own, or regularly help to look after, a horse or pony, is capable of learning to effect simple saddlery repairs. Then, with more experience, they can progress to more complicated jobs.

Secondly, and it hurts to say this, I have seen many repairs carried out by 'saddlers' that are, to say the least, shoddy. In many garages today a lot of mechanics are only fitters; they could not strip and repair a starter motor, only fit a factory reconditioned one. Similarly, most saddlery shops trading today base their business on new sales of bought-in saddlery rather than manufacturing it themselves. As one leading retailer put it, 'Tell the customer that it can't be repaired, and immediately introduce a new item and try to make a sale. It's a lot less work'. Yes, it is less work for him, buying as he does from a factory, but what is this doing for the craft of saddlery, not to mention your pocket?

Please read these chapters in order because each will expand on the tools and their uses chronologically, and show how with practise and, above all, patience you will be able to carry out prompt, neat, and, most importantly, safe repairs to your saddlery.

1

TOOLS AND MATERIALS

The old saying 'A bad workman always blames his tools' can equally be set against, 'Give a workman the tools to match his skill'. As with any trade or craft, certain tools are absolutely essential, and without which a start cannot be made. Once you have gained some experience and acquired a certain amount of skill, you can progress further, adding the required tools as you go along (see Fig. 1.1).

I have divided the tools up into three sections: beginner, improver and experienced.

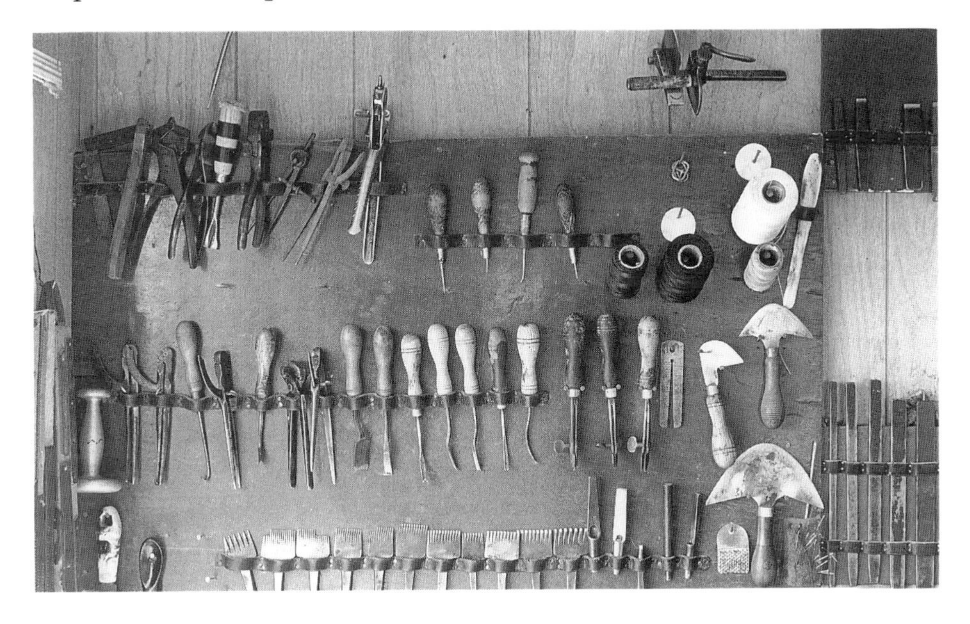

Fig. 1.1 Part of the author's tool collection. With the tools mounted on a board in this way, it is easy to select the appropriate tool as needed.

Suggested Tools for the Beginner

Stitching clamps

A three-piece pair approximately 33 in high would be an ideal pair to start with.

Awl

You will need one or two handles and several spare blades 2–2½ in long because you will be certain to snap some when beginning to stitch. (See note at the end of the tool section on how to change blades.)

Wheel punch

A punch with a revolving head and six different sized holes will be invaluable.

Knife

If you cannot buy a round knife to start with, it is surprising what you can do with a Stanley knife.

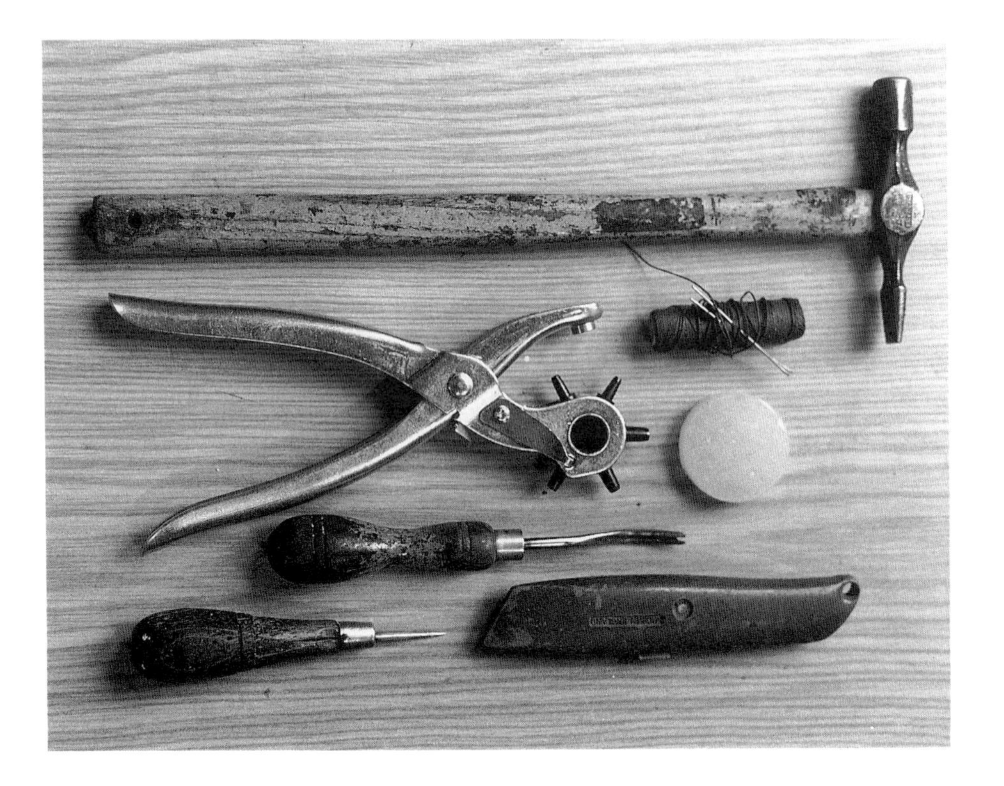

Fig. 1.2 Many basic repairs can be effected with the beginner's basic tools.

Needles

A packet each of sizes 0 and 3 will suffice. Instructions for making curved needles are given later on in this book.

Hammer

You will ideally need two; a lightweight tack hammer and a claw-or ball-head type with a nice large flat face.

Edge tool

Buy one if you possibly can once you start to use new leather for repairs because it chamfers the edge of the leather. Initially, buy a No. 1 or No. 2.

Extra Tools for the Improver

This list is for improvers who have mastered hand-stitching and simple cutting out and are now ready to try their hand at more elaborate repairs.

Edge tool

After the No. 1 or 2, buy a No. 4 or 5 for the heavier work.

Pricking iron or stitch marker

To start with, buy irons covering eight or nine stitches to the inch for bridlework and six or seven to the inch for headcollars and girths etc. The size will be marked on the iron.

Crew hole punch

This is an oval punch used to cut a hole to take the tongue of the buckle. They range in size from 34 to 40. The two most useful sizes are 36 and 39.

Round knife

The round knife is the proper saddlers' knife used for cutting and skiving leather. Do not get anything too large; 5 in is ideal to start with.

Loop sticks

Loop sticks are made from wood or metal; they are used to block the leather loops for a buckle to size and shape. Sizes range from

3/8 in to 1¼ in. To start with you could make your own from wood approximately 10 in long and ¼ in thick. Smooth the edges with glass paper so they will not cut the inside of the leather loops.

Fig. 1.3 Once you progress to using new leather, you will need a few additional tools to facilitate cutting, marking out and finishing the work.

Creasing iron
This is a double-bladed tool which, when heated, is drawn along the length of the leather giving it that distinctive dark edge line. A screw allows for adjustment of the blades.

Compass
With a screw to lock its position, the compass assists with the marking out of holes. If you need to cut a straight piece of leather off a hide and do not have a plough gauge, then this tool will give you an accurate measurement to cut along.

Oval hole punch
While holes for light bridles and headcollar buckles can be made with a wheel punch, the heavier buckles of stirrup leathers, girths and driving harness really should have an oval hole for the tongue to lie in.

Bone folder
This tool is useful but not essential. When rubbed hard and fast along the edges of leather which have been stitched together, it smooths and evens out the surface.

Lead block
All punches that cut out leather, i.e. crew hole and oval hole punches, must be used against something soft, such as a lead block, or the sharp edge of the blade will be damaged. A block of end grain of wood could be used as a substitute for a lead block, but this cannot be smoothed out when worn.

A weight, approximately 28 lb (12.7 k)
An old shop-scales' weight or anything heavy with a flat surface will come in very handy in the workshop.

Oilstone
An oilstone is needed to sharpen the saddlers' round knife. If you have a two-sided stone, start with the coarse side and use a circular movement while holding the blade at a low angle, turning it over regularly. To put a finishing edge on the knife, raise the blade to a slightly greater angle, and use gentle strokes on the fine side of the oilstone (see Fig. 1.4). With a single-texture stone

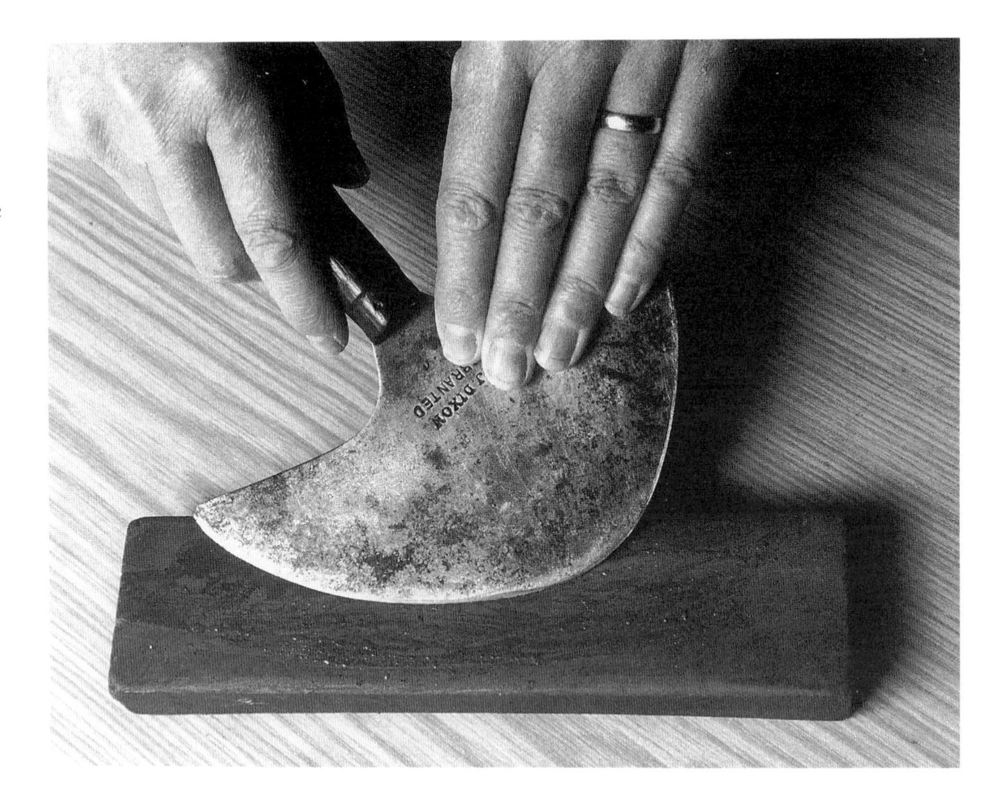

Fig. 1.4 To keep a keen edge on the saddlers' round knife and head knife, sharpen them regularly on an oilstone. Draw the knife across the stone at an angle of 45 degrees using a circular motion.

simply alter the angle of the blade and the type of strokes to sharpen the knife.

Most of the repairs in this book can be accomplished with the tools already listed. The following selection of tools would also enable someone experienced in most repairs to try their hand at manufacturing new or replacement parts.

Additional Tools for an Experienced Worker

Edge tool

If you do not yet have a No. 4 or No. 5 edge tool, then now is the time to get one.

Crew hole punches

Add to your collection with numbers 35 and 37.

Smasher or masher

The smasher is not dissimilar in shape to the tool that is used to facilitate the darning of socks. It is available in traditional wood or modern aluminium which I much prefer. Although used by the saddler when restuffing saddles, it is also a useful tool when relining driving collars and saddles (see Fig. 1.5).

Fig. 1.5 These extra tools are only needed if you intend to use new leather regularly and tackle complicated saddle and collar repairs.

Half round, or head, knife

Basically, this is half a round knife and, while not essential, I find it very useful when working on saddles because it is possible to get the blade into tight corners.

Leather thickness gauge

This gauge is a handy little tool for calculating quickly the thickness of a hide (see Fig. 1.6).

Plough gauge and knife

The easiest way to cut uniform strips from a hide is with this instrument. A calibrated bar allows a wide range of sizes to be cut at ease (see Fig. 1.7).

Fig. 1.6 The thickness of a hide is measured in millimetres and this can be easily ascertained by using this simple, but effective, gauge. This hide is just under 4½ mm thick.

Fig. 1.7 The plough gauge has calibrated markings on a bar along which a guide slides to enable strips of leather to be cut with ease. With this tool you will be able to adjust the width of the leather to be cut, from ⅜ in to about 4 in. Periodically you will need to remove its blade-cum-handle and put a fresh edge on it with an oilstone.

Splitting machine

This machine has an 8–10 in blade with a spring-loaded brass roller that holds the leather against the blade, allowing the leather to be pulled through, splitting the thickness in the process (see Fig. 1.8).

Fig. 1.8 The thickness of a piece of leather is reduced by pulling it through this splitting machine between the knife edge and the spring-loaded roller. A wheel at the base of the machine adjusts the height of the roller thus governing the thickness of the split leather.

Sewing machine

A sewing machine is invaluable for rug repairs as well as repairs to web headcollars and harness. All the repairs in this book can, however, be carried out without a sewing machine.

These lists are only a suggested collection for each stage of competence. Very often you will find tools for sale at auctions, either individually or as an assorted lot. The Reading auctioneers, Thimbleby and Shorland (see Appendix) regularly offer for sale such lots at their Carriage and Harness Sales.

Other sources of tools are also given at the end of this book, or you could be lucky and buy a ready-made collection. When I started my apprenticeship I bought a box of assorted tools from a leather worker who was retiring. He had done just the same thing when he had started his working life some 50 years before, and some of those original tools were the ones he sold to me. They are still in my workshop and still in use.

Replacing Awl Blades

As mentioned before, the beginner is sure to snap an awl blade or two in the early days, so I shall explain the best way I have found to replace a broken blade.

If the blade has broken flush with the handle you will not be able to remove it, so a small pilot hole must be drilled or tapped into the handle beside the old blade to accommodate the new blade. If part of the blade is still standing out from the awl take two two-pence pieces (or any other coinage made from soft metal) place one each side of the blade, insert into a vice and screw as tight as possible. Then pull the awl handle away sharply; the blade should come out of the handle. If the blade merely slips out from between the two coins, repeat the procedure, tightening the vice a little more. When the old blade is out, place the new awl blade in the old or new pilot hole, again, protecting the blade with the soft metal coins, and place in the vice. Then gently tap the end of the handle with a hammer until the blade is in to the required depth.

Materials

Over the last few years considerable research has gone into the manufacture of new materials and products, some of which have made their way into the saddlery trade. Waterproof New Zealand rug manufacturers have made full use of the new cloths available to them, plastic (synthetic) saddles have, after a shaky start, now established themselves, and webbing exercise driving harness is almost the norm nowadays. Some of these specialist materials are hard to obtain in small quantities for repair work and alternatives must be sought.

Apart from leather, which I will deal with later in this chapter, the materials most likely to be needed to repair saddlery and harness are the following (their sources are listed at the back of the book in the Appendix).

Canvas webbing

This is for the repair of rugs. Heavy-quality denim cloth can also be of use for small repairs.

Rugging
Never throw away any old blankets because they can always be used to line rugs, rollers or collars.

Felt
Reflocking a saddle is a highly skilled job, but foam or felt fillets can be substituted.

Nylon webbing
Used for repairing stable rollers and driving harness. Nylon webbing can also replace the leather fittings when repairing rugs.

Beeswax
Invaluable for waterproofing stitching thread but a candle can be substituted if need be.

Thread
For most of the repairs covered in this book, ordinary linen plaiting thread will suffice. For the heavier jobs use it double thickness.

Copydex
The best known latex adhesive. Unbeatable for rug repairs because it penetrates the weave of the material.

Leather glue
My favourite brand is Thixofix made by Dunlop because it spreads like butter and is not stringy like many others. When gluing leather to leather a solvent-based glue is stronger than Copydex latex adhesive.

Dye stain
A water-based dye solution is used to help seal the edges of leather, whereas a spirit dye is useful to darken the top surface of leather.

Rubber rein grips
Nowadays available in a variety of colours, these grips still make the best wet-weather reins.

Velvet ribbon
Coloured velvet browbands are as popular as ever for pony show

bridles, and a couple of metres of ribbon can transform a bridle.

Buckles, clips, rings and fittings

Basically, there are eight different types of materials used in the manufacture of these items. The first six are metal or metal based, the last two are synthetic.

BRASS

This metal looks good when it has just been polished, but tarnishes quickly. Although totally rust proof it is not a strong metal in comparison to stainless steel.

STAINLESS STEEL

Definitely the strongest metal for manufacturing buckles, stirrup irons and bits. Contrary to this metal's description, however, I find that stainless steel stirrup irons will, if left wet, actually rust. For girth and stirrup leather buckles, stainless steel is, without doubt, the best material.

SOLID NICKEL

A non-rusting metal used for some bridle and stirrup leather buckles and hookstuds on bridle billets.

NICKEL-PLATED STEEL

A cheaper version of the above, but once the plating chips or wears off, the steel underneath is vulnerable to rust. Many trigger clips are made from this metal, so always keep the moving parts of the clips well oiled.

ZINC-PLATED STEEL

Fittings, mainly rings, made from this metal are meant to be rust-proof, but the last batch of rings I had rusted almost overnight. They are, however, very strong.

COLOURED, NON-RUST COATINGS

Normally black, to blend in with web driving harness. They look good when new, but the coating will chip when the harness is in use. It is the solid metal inside the fittings that gives them the strength.

NYLON AND PLASTIC

Always view with suspicion, because, although plastic and nylon have come a long way, I have seem many headcollar and surcingle fittings made from synthetic materials break off all too easily!

Measuring Leather

In Britain we use metric measurements for the thickness of a hide and imperial measurements for the surface area. Therefore, you may buy a hide 3 mm thick × 10 sq ft. This may seem strange, but it is a tradition. On the Continent the surface area is measured in square metres.

Bridle butts and harness backs are sometimes sold priced as a pair irrespective of their size.

Hide thickness

2 mm

Used only for items that do not come under much stress and need to be very supple, e.g. spur straps and straps on brushing boots.

3–3½ mm

The ideal bridle-making weight of leather. Kept oiled, this weight will be supple enough to use on narrow buckles and will bend to go through loops or keepers. This weight can also be used on pony headcollars and surcingle straps.

4–4½ mm

The heavyweight end of saddlery leather. This is used mainly for stirrup leathers, rug fittings, larger headcollars, girth straps and Balding girths. It is not normally cut under ¾ in wide because it would be difficult to use on smaller buckles. Because of its thickness and generally harder substance, when using this leather you need to take a greater amount of its edge off when preparing it in order to achieve a rounded edge. This gives a neater appearance and contributes to the horse's comfort.

For Balding girths and stirrup leathers I normally use a No. 5 or No. 6 edge tool.

Types of Leather

Bridle butt

Measuring approximately 5 ft × 2 ft, this is the most expensive

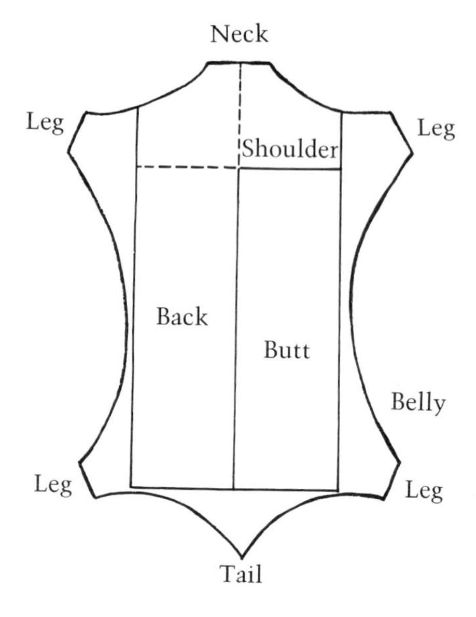

Fig. 1.9 This diagram shows a trimmed hide. The leg, tail and belly parts are not normally used in saddlery owing to their poor quality. The hide can be cut either into two backs which would include part of the shoulder and give greater length, or into a pair of butts together with the shoulder section. (Not to scale.)

part of the hide, and the best. The weaker belly part is trimmed off, as is the coarser shoulder. The bridle butt is often sold as a pair, because it is actually the two choice sections taken from either side of the central spine or middle part of the animal.

Harness or bridle back

Once again the belly part has been trimmed off the hide, but extra length is obtained by including the shoulder part. This hide is especially useful when making a set of harness because it gives you the length to cut traces, backbands and reins. Care has to be taken with this type of hide because you seldom get uniformity of quality along its length.

Shoulder

Although often used in the manufacture of second-class bridle work, this part of the hide can be very useful for repairs.

Belly

This part of the hide was the most flexible whilst on the animal, so its loose texture and poor strength make it generally unsuitable for use within the saddlery trade. It can be used as backing for other pieces of leather, but should never be used where it will come under any stress.

Printed shoulders and butt

Many years ago the flaps, skirts and seat of a saddle were made from pigskin; the flaps and skirts being backed onto a heavier hide. Subsequently, for economic reasons, the flaps and skirt were cut from a solid cowhide, but with the distinctive texture of pigskin embossed on to it. This produced a good-looking effect that was far harder wearing than the true pigskin-covered saddle. Some quality foreign makers, notably German, even embossed a heavier pattern on to their flap leather to aid grip.

Nowadays many manufacturers prefer a smooth oiled hide for use on skirts and flaps, together with hides other than pigskin for the seat.

Chrome leather

Grey in colour and tanned to be water resistant, it is normally only used for New Zealand rug leg, surcingle and breast fittings, and occasionally for stable headcollars and girth straps. This leather is usually available in ready-cut strips, but can be replaced by other materials for repair work.

Buffalo hide

Dark maroon in colour and unbreakable in use, buffalo hide's only disadvantage is that it stretches a great deal and cannot be finished as neatly as other cowhides. Buffalo hide, sometimes incorrectly called rawhide, is used mainly for high quality stirrup leathers and girth straps. As with chrome leather, buffalo hide normally comes in ready-cut strips, but because it is unbreakable should never need replacing. My own pair of 1¼ in wide stirrup leathers with stainless steel buckles have been in use for over 10 years and are still in first class condition.

All the above hides range in thickness from 2 mm to 5 mm and are used for strapping and saddle flaps. We now come to the lighter skins which are generally used as coverings.

Basil, kip or goat

These hides vary enormously in quality and colour. They are only to be used when no great strain is to be applied, e.g. a cheap overline on an old surge-lined saddle.

Lining or panel hide

This cowhide is available in either a dry or, as I prefer, a slightly oily waxed finish. It is traditionally used to line saddles and collars. Although not cheap, it is, if you can afford it, a far better choice than the previous hide.

Bag or coach hide

A really tough hide which was, as its name suggests, used in the manufacture of Gladstone and money bags as well as hoods for driving vehicles. Today it is usually employed for three-fold girths and some brushing boots, if the hide is not too heavy. I have re-covered large swept wings (mud guards) from an antique Victoria phaeton with this hide; it was good to work with and looked authentic.

Patent hide

This is normally black, but is available in shades of red, yellow, blue and green. This is the shiny leather you see on show driving harness and on the vehicle's shafts, dashboard and wings. With trade harness, that is harness to be used in conjunction with a commercial vehicle, i.e. costermongers or delivery service, you will often see key parts of the harness trimmed with a bright colour. This is sometimes called gypsy harness. Not all of this is leather because nowadays plastic is sometimes used. True patent is a coloured finish sprayed onto a leather backing. The main difference, apart from price, is that, when soaked in warm water, the leather will stretch slightly to the required shape. Despite this advantage, patent must be about the least hard-wearing leather available because even a slight scratch will be very obvious. For this reason patent is used mainly on show harness and vehicles where it will only be used lightly for short periods of time.

Pigskin

Today, pigskin is used only for saddle seats, piping on the panel and sometimes padded knee grips. This hide is immensely strong because it has a very dense structure, and is easily recognized by the tiny bristle holes in groups of three. When inspected closely you will see that these holes penetrate the whole thickness of the skin. In my opinion this is still the finest material with which to make a saddle seat.

2

GETTING STARTED

Having talked about leather and the tools needed to work it, we now move on to building your confidence and, consequently, your skill, to tackle more complex tasks. All too often old stable headcollars, stirrup leathers and bridle work are seen repaired with a rivet; sometimes the rivet may outlive the leather work, but it is unsightly and often unsafe. Leather can stretch, rivets can work loose, and if a rivet has a rough edge this could easily rub a sore on the horse. Forget using rivets and learn to effect a good safe repair using traditional materials.

One of the easiest ways to get started with your own repairs is to tackle restitching jobs which utilize existing stitch holes and do not require any new leather or fittings. These jobs include stirrup leathers, rug fittings, girth straps, the ends of saddle panels, and padded knee grips on saddle flaps. All of these regularly require attention owing to the friction imposed upon them in use.

Let us start with the ends of the saddle panel. No matter how often you oil the panel the stitching will always wear through where the girth lies across them (see Fig. 2.1). It is not a matter of safety, but, if this job is not attended to promptly, the leather starts to curl and break up, making it necessary to patch the ends with additional pieces of leather.

The first job is to pick out the old stitching which will normally wear from the girth strap side, so hook the stitches out on the reverse side. You should be left with nice large holes to work with. It may not be necessary to restitch the entire length of the

Fig. 2.1 The ends of the panel will always wear at the point of contact with the girth and it is a straightforward job to restitch using one needle. You simply need to pick out the old worn stitching, follow the original stitch holes with an awl, and restitch with waxed thread. Clamps are not required.

panel, but only five or six inches where the damage has been done.

The next job is to prepare the thread. If using ordinary plaiting thread, cut a piece measuring two arms' length and use doubled. To waterproof the thread and help prevent it from fraying when stitching, draw it through a block of beeswax several times, this will make it slightly tacky and help preserve the thread in use (see Fig. 2.2).

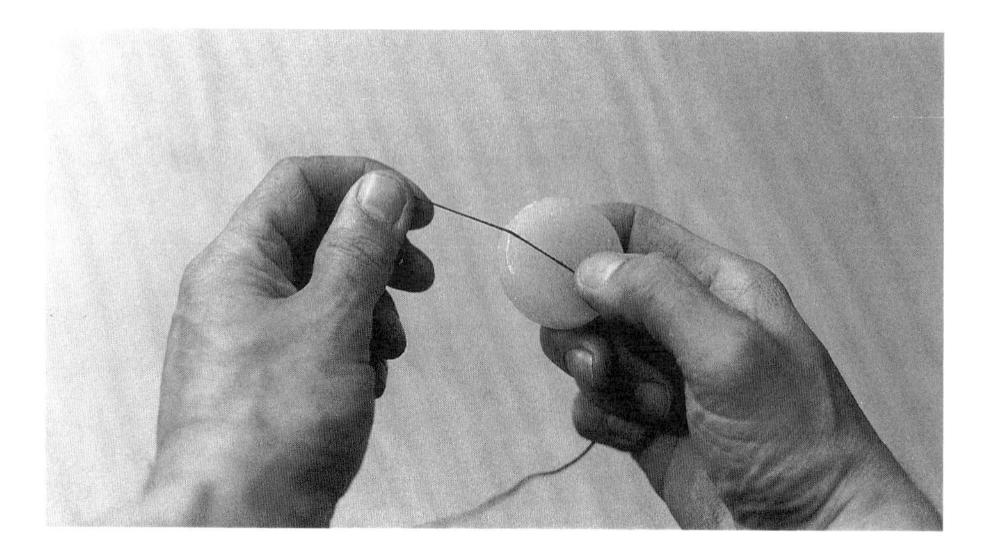

Fig. 2.2 The linen thread used for stitching is first drawn through a piece of beeswax which helps to protect it. Rain, horse sweat and abrasion all cause stitching to wear, and a coating of wax can greatly increase its life. During stitching you may find that the thread starts to fray or come apart, if this happens rewax it either with the needles left in place or unthreaded.

Thread a blunt-ended needle, and knot the ends together. Now push the needle through the vacant holes first making a stitch then missing a stitch (pop stitching). If the holes are large enough you may not need a stitching awl, especially for the first row of stitches. When you have reached the part where the old stitching is sound, overlap the new stitching for an inch or so then work

Fig. 2.3 When stitching, always make use of the awl handle to push the needle through the work.

back, in and out, so completing the stitches you missed before. Because the holes now have some thread running through them, you may well need to use a stitching awl; simply push the awl through to make way for the needle. When you have finished, go back over the new stitching for a few stitches then cut your thread. There is no need to tie a knot to finish and it is best to cut off the knot you started with if it stands proud of the surface, especially if it is on the horse side of the panel.

Lay the ends of the panels on a smooth flat surface and gently flatten the stitching with the widest-faced hammer you have. This will lay the stitches, and reduce the points of wear when used in the future.

This job can also be effected with two needles, (two-needle stitching is described later) but for the beginner one needle will suffice. If the job consists of only four or five inches of stitching I would normally use just the one needle myself. This job does not require stitching clamps because it is easier to work on the saddle on a work-bench top.

Knee pads can also be repaired in this way, although, because you will be making finer stitches, only a single thread is required,

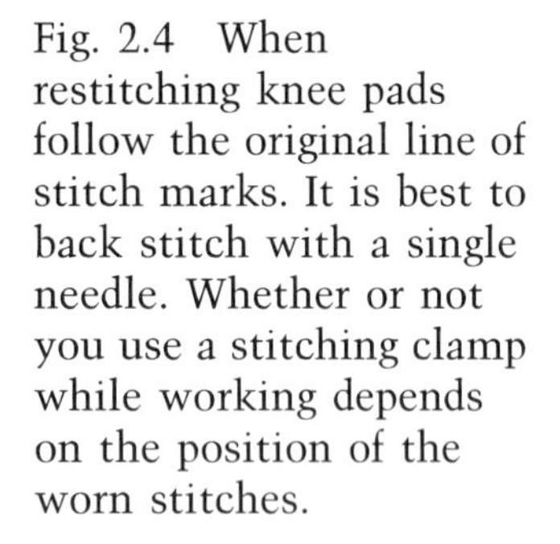

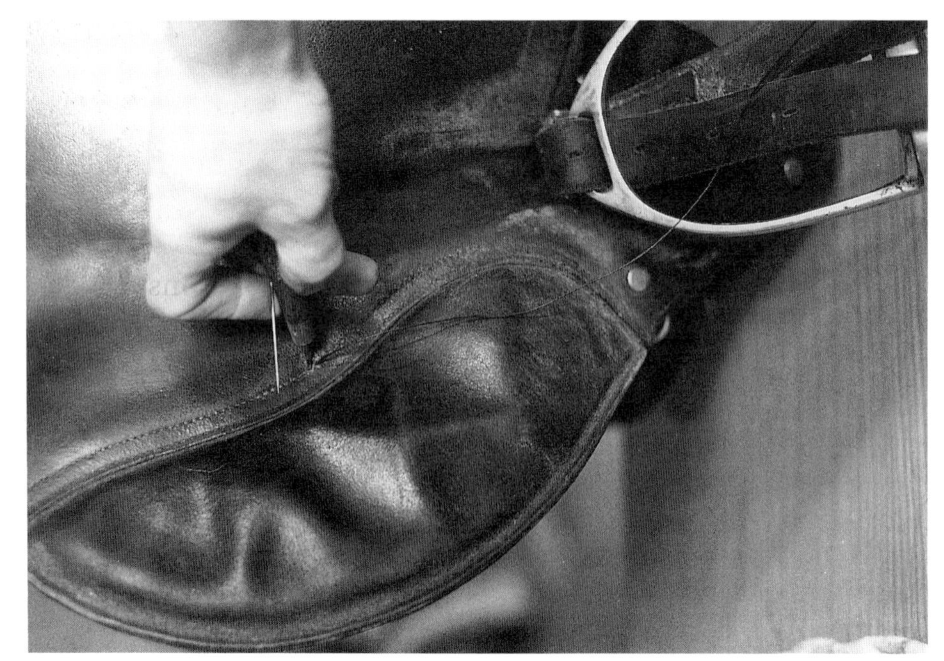

Fig. 2.4 When restitching knee pads follow the original line of stitch marks. It is best to back stitch with a single needle. Whether or not you use a stitching clamp while working depends on the position of the worn stitches.

and you will most certainly need to use a stitching awl (see Fig. 2.4). It may also help to use a pair of clamps to hold the saddle flap, and you will need to juggle the saddle around to find the best and most convenient position for you to stitch it. (If there is a second pair of hands available, ask for help in holding the saddle in a certain position while you stitch).

Repairing knee pads looks straight forward, but the saddle can rock and fall about if not held securely – rather like a set of bagpipes in the hands of a novice!

Stitching Clamps

Stitching clamps are one of the most indispensable pieces of equipment in the saddler's workshop. They are made from two pieces of wood, steam bent to give them some spring when in use. Basically, these hold pieces of work firmly in place in front of the saddler for stitching. In my workshop we use the tall, 3½ ft high clamps. A length of leather is attached to one side of one of the clamp arms, passes through the other arm, and hangs down with a stirrup iron at the end (see Fig. 2.5). When using tall clamps, a high-seated chair is required. The stitcher works with the clamps held upright between the legs, and with a foot in the stirrup applying pressure to hold the work tight. By releasing the pressure on the stirrup the work can be taken out or adjusted, then held tight again once pressure is reapplied.

The simple-type clamps are approximately 2½ ft high and can be used while seated in a standard-height chair. They are usually made from three pieces of wood (i.e. the two clamp arms and the middle section that stands on the ground), so it is an easy job to saw the middle piece down to the required height. These clamps do not employ a stirrup iron and leather, so slightly more knee pressure is needed to hold the work securely in place. These are the clamps I would recommend for the beginner because they are easier to handle, can be operated from an ordinary chair and cost less.

I have a two-piece pair of steam-bent clamps, approximately 2½ ft high, which I always used when doing stitching demon-

Fig. 2.5 The saddler's stitching clamps are extremely useful. This pair have the added advantage of a stirrup iron and leather to help keep the tension on the clamps while stitching; this saves the legs having to squeeze the clamps together.

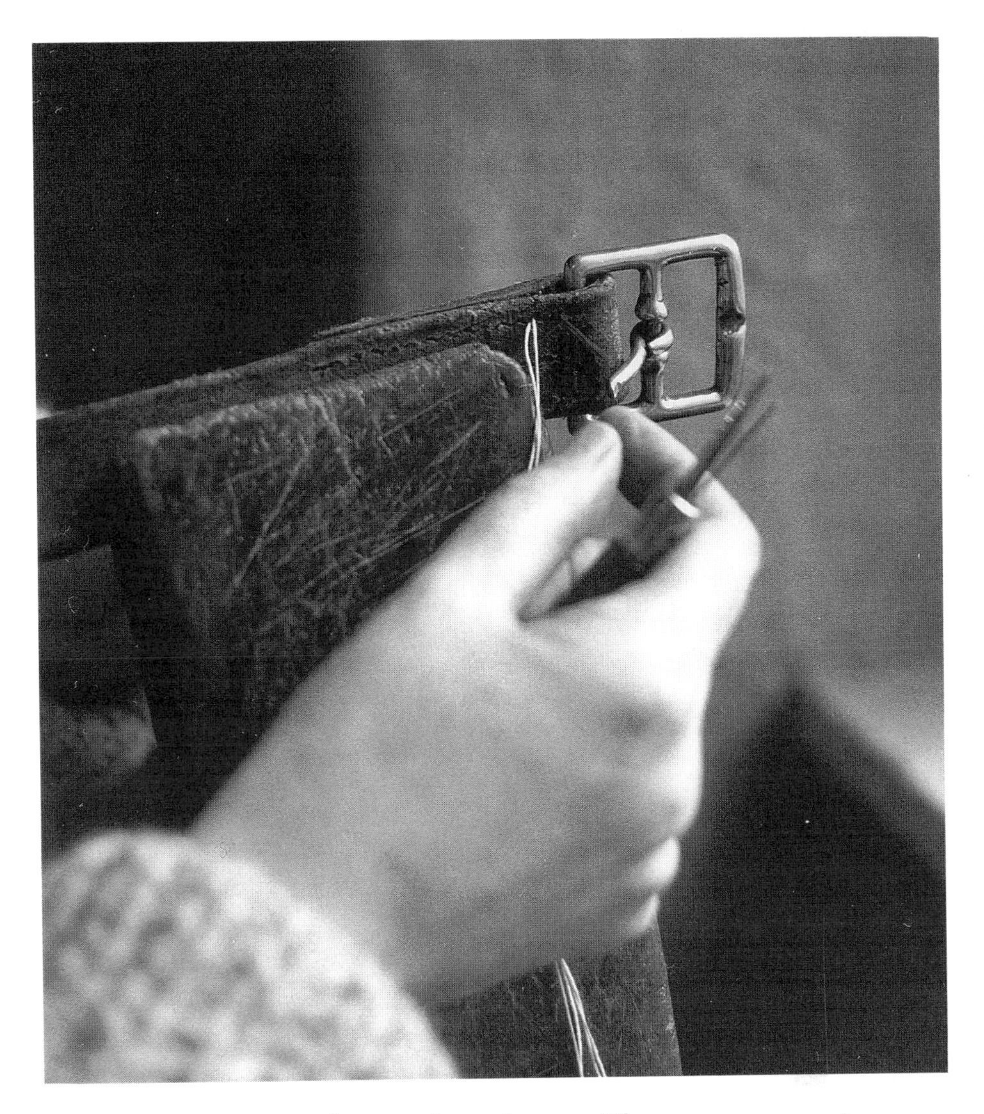

Fig. 2.6 The piece of saddlery to be stitched is held between the two faces of the clamps with the line of stitching just showing. This arrangement holds the work securely in place leaving both hands free to stitch. I know people who have adapted an ordinary woodworking vice to hold work for stitching, and, evidently, it can work quite well occasionally. If, however, you are going to stitch on a regular basis, I would strongly recommend buying a pair of proper clamps.

strations on my trade stand at shows. They now come in very handy for stitching in the garden during the warm weather.

When holding work in the clamps, adjust the work so that the piece to be stitched runs horizontally with the top face of the clamps and about ¼ in or so above it (see Fig. 2.6). It is obvious that the narrower the width and finer the substance of the leather you are holding in the clamps, the harder it is to work on. Hence it is easier to hold a 1¼ in wide hunting stirrup leather than a ⅜ in wide rein from a pony show bridle.

Work method when using clamps

When stitching leather held in clamps, you always work towards you, in other words, you start stitching at the end furthest away from you. Therefore stitch hole number one is the hole furthest away from you.

If you are right-handed, have the top-facing side of the work on your right with the stitches marked on the right side. If left-handed, have the work facing left.

Double, or two-needle, stitching

When learning to use the clamps, an ideal job to start with is the restitching of stirrup leathers; because these have no leather loops (keepers), you can concentrate on the stitching technique. Use two needles; you could use one, but the correct way to achieve a good strong repair is with two needles. Cut your thread. You will get to know how much thread is needed for a job, but, as a general rule, if you hold the reel of thread in one hand and draw the end across your body till both your arms are outstretched, this will be more than enough for you to cope with in the beginning. Now wax the thread, and then thread a needle onto each end, making sure the return of the thread is the same on each needle. Because you are going to stitch using the middle of the thread, you will not need to tie any knots.

Many people think that just because the last three or four stitches of a stirrup leather have worn away, these are the only ones needing restitching. This is a dangerous assumption to make. Very often, if you hold the end of the return on the stirrup leather and pull sharply, most of the other stitches will break (see Figs. 2.7a and b). So, for safety's sake, always cut through any other stitches and restitch the entire leather.

Having picked out all the old stitches from the stirrup leather with an awl (see Fig. 2.8), square up the stitch holes by placing the awl in the end stitch hole on both pieces of leather (see Fig. 2.9), then place it in the clamps. Now push a needle through the second hole thereby having a needle on each side of the work. Hold the needles up to the full extent of the thread and ensure the needles are level. If the returns of the thread are also equal you will have the same amount of thread to stitch with on

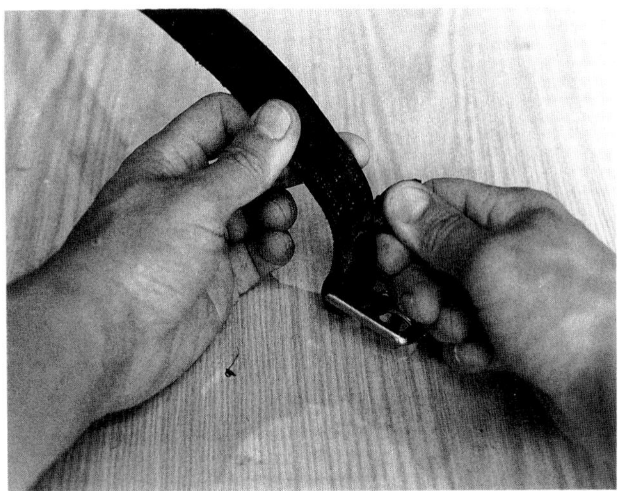

Figs. 2.7 a and b Although it may appear that only the first three or four stitches require repair, it is good, safe workshop practice to restitch the entire stirrup leather.

each side of the work.

Unlike machine stitching where one thread runs along the top and another underneath, with two-needle, or double, stitching each thread will alternate each side of the work. In effect the threads cross themselves each time they make a stitch and this is what gives double stitching its great strength. If one stitch

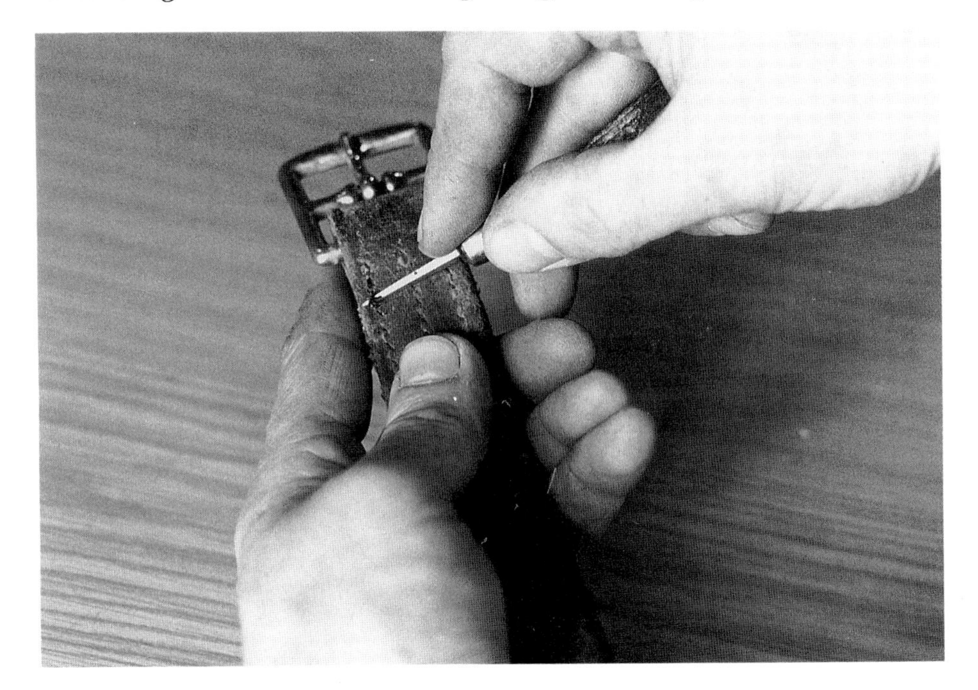

Fig. 2.8 Before commencing to restitch any item of saddlery, always pick out the old stitches with an awl. This not only results in a neater job, but also makes stitching much easier because the holes are all preformed.

Fig. 2.9 Once the old stitches have been removed, match up the top holes in each side of the stirrup leather with an awl as shown. The work can now be put in the clamps ready for stitching.

breaks then the others should take the strain, unlike machine stitching when if one stitch breaks the rest can run very easily. When this is understood, it is obvious how dangerous it is to machine stitch the parts of saddlery that take great stress.

Starting with the left needle, pass it to the right, through the end stitch hole, then pass the other needle from right to left through the same end stitch hole. Make the first stitch twice for added strength. Having made your first stitch, start to stitch towards you. Push the awl through first each time to line up the holes, then pass the left needle through followed by the right needle, continuing to work towards you. As each stitch is made, pull evenly and firmly on each needle at the same time, this will draw the leather tightly together, but there is no need to exert a trial of strength on them. At the end of the row push the right-

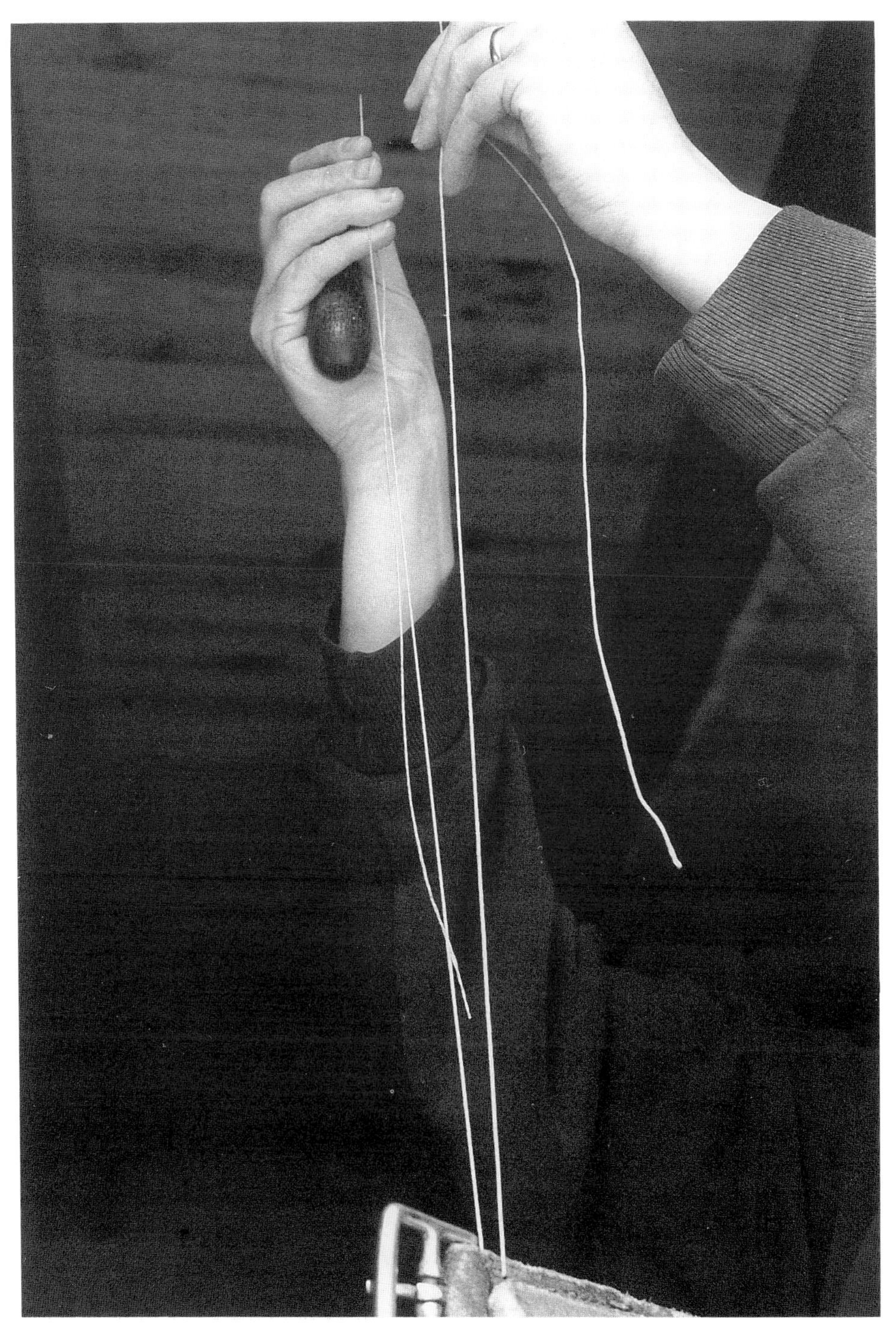

Fig. 2.10 Having threaded two needles always make sure that the return of thread on each needle is equal. When you have passed the thread through the first stitch hole, hold the needles up together to centralise them. This way you will ensure you have an equal amount of thread each side of the work.

hand needle back one hole so that both needles are on the left-hand side.

Take the work out of the clamps and turn it round to stitch the other side. Your needles will now be at the far end of the work, twist them around each other two or three times in order to keep the thread together, pull gently and push one needle through to the right. Now continue, stitching towards you just as before, until you have completed the last stitch, then go over the last two stitches again. Finish with both needles on the left-hand side. There is no need to tie a knot because you have overstitched the last two or three stitches, so simply cut the threads as close to the leather as you can. As before, lie the stitching on a flat surface and gently tap with a broad-faced hammer to help sink the stitches flush with the leather.

Single-needle, or back, stitching

With this method of stitching, thread the waxed thread onto one needle and tie a neat knot at the end. On the front of your work (the side that is normally seen when the equipment is in use) the stitches will be the same size as in double stitching, but

Fig. 2.11 In this picture you can see quite plainly how, when using only one needle, the length of the stitch on the under, or left-hand, side will be twice that on the top, or right-hand, side. This is assuming the stitcher is right-handed. If the large stitch at the back of the work is always lifted above the needle as it passes through to the left side, the stitches will lie at a neat angle.

the length of the stitch at the back of the work will be twice as long. For many jobs this would be unsightly, and, because of the greater length of stitch, would wear more quickly in use, therefore it should never be used on headcollars, or stirrup leathers. It does, however, come into its own when stitching leather onto web or rugging because the large stitch behind helps to grip the weave of the material, therefore lampwick girths, rug and roller fittings, and girth straps on saddles are more secure when stitched in this fashion.

With the work in the clamps, the awl is pushed through the second stitch hole and the needle drawn through (left to right) leaving the knot on the back surface of the work. Now put the awl through the first stitch hole, then pass the needle through to the same side as the knot. After putting the awl through the third hole from the end, pass the needle through to the front of the work and return it to the back through the second stitch hole.

You will now have two small stitches on the front side of the work and the stitch on the back of the work, twice the size. (See Fig. 2.11 which shows exaggerated-size stitches to emphasize the point.) When the work is completed finish off the last stitch simply by tying a single neat knot as close to the back of the work as possible.

3

BRIDLE WORK

The term 'bridle work' covers not only the specific making and repairing of bridles, but also the making and repairing of points (straps) on rugs, surcingles etc., and the preparation of the hide for this work.

Hide Preparation

Once you have some experience of stitching with both one and two needles, restitching straightforward jobs such as girths, stirrup leathers and headcollars, it's time to progress towards cutting out leather replacements. Never throw away leather reins or similar pieces because the sound parts can always be used for repairs. This has the added advantage of making the repairs blend in with the bridle or piece of equipment. I do not like repairs that stand out from a piece of saddlery; repairs should be sympathetic to the original as well as safe.

The photographs of the hunting horn case are a good example of this. In Fig. 3.1 the replacement strap at the top fulfils the job of holding the horn case to the saddle, but hardly blends in with the rest of the well-used case and saddle. By using darker leather, staining to match or using an old, but sound, part of a rein, this repair is quite different (see Fig. 3.2). The same criteria can be applied to bridle cheek pieces and many other parts of saddlery (see Fig. 3.3).

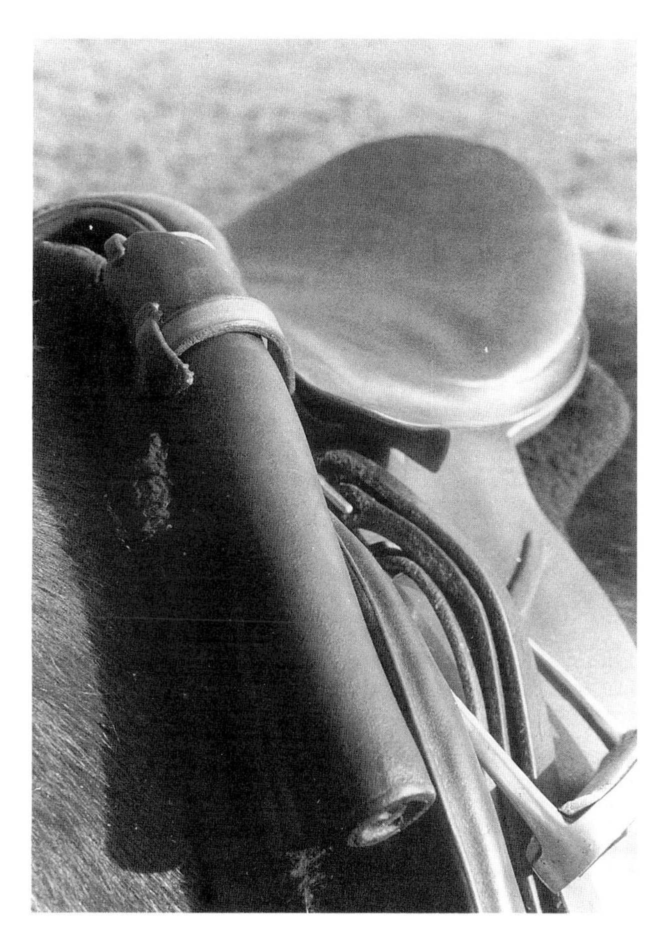

Fig. 3.1 On this horn case repair, a lighter piece of leather has been used to replace the top strap.

Fig. 3.2 It is much neater if either a darker stain of leather is used, or a sound piece of old leather, such as a rein, is cut to size. Always try to match the repair to the original piece of saddlery.

The leather of a lovely old brown set of driving harness is, in my experience, the hardest leather to match when executing repairs. Over the years the colour of the leather has matured to a mahogany-like finish. The best way to deal with this is by careful application of brown shoe dye, wiping it off before it dries completely, so effecting an uneven colour to match the rest of the weathered harness. The original work may well have been stitched with white or yellow thread, but when repairing the harness try to match the *existing* colour of the stitches. This may mean using either brown or black thread.

Let us now look at some of the simpler repair jobs which require leather replacements.

In Chapter 1, I discussed the different types of leather and the

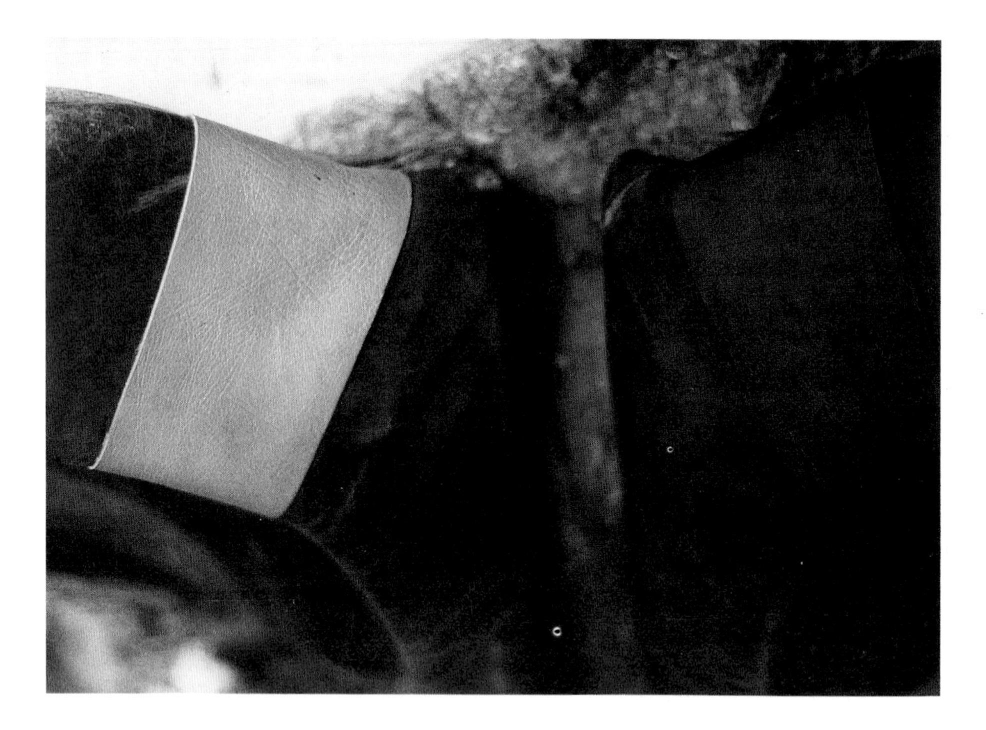

Fig. 3.3 Even a patch-type repair needs to blend in with the piece of saddlery or harness. The difference a small amount of leather dye can make is shown in this picture. On the left the patch stands out badly from the rest of the panel, whereas the dyed patch on the right blends in nicely with the saddle.

use for which each was best suited, but if I were allowed only one hide to use on repairs, I would choose a 3½ mm shoulder, or, if funds permitted, a bridle butt in a mid-brown colour. Brown suits most bridle work and could be stained if need be. At 3½ mm thick it would be ideal for replacing rein billets, surcingle straps, and for headcollar repairs.

Points (straps)

Initially, let us deal with a new point to go on a surcingle or roller. The internal width of the buckle will give you the width required to cut the point, usually either 1 in or 1¼ in.

With the hide on the bench the first job is to cut a straight edge from which you can work. Using a long piece of wood with a least one true edge as a guide, mark along the length of the hide as near to the edge as possible. Now trim along this line using a round knife, if possible, but a Stanley-type knife will suffice (see Figs. 3.4 and 3.5).

Fig. 3.4 The correct way to hold a piece of leather while cutting with a saddlers' round knife.

Fig. 3.5 The incorrect way to hold a piece of leather. One slip of the knife and your fingers will be endangered.

Next, set your plough gauge to the required width, or failing that, mark the width directly onto the hide with a compass. When using a plough gauge, you pull the strip to be cut through the plough gauge with your left hand, while pushing the plough forward with your right (see Fig. 3.6). At all times, keep the edge of the hide against the guide.

Fig. 3.6 Although you can cut a strip of leather off a hide without a plough gauge, it is the correct tool for the job. The sliding gauge on the left allows you to cut strips in eighth-inch stages. A second gauge in the form of a brass roller can be adjusted to suit various thicknesses of leather.

Never try to cut off only the required length with a plough gauge, leaving part of a strip uncut on the hide. Although this will do for the first job, cutting subsequent widths will be far more difficult. Cut the entire strip from the hide, trim to the amount needed for the current job, and store the unused leather for future use.

If you do not have a plough gauge, it is quite possible to cut the strip off the hide with a hand knife.

Next, measure the length of the point, (e.g. 16 in) then cut to size and shape at one end. This can be done either with a pointed hand punch, or marked as a church window, or as a straight point, and cut with a knife (see Fig. 3.7).

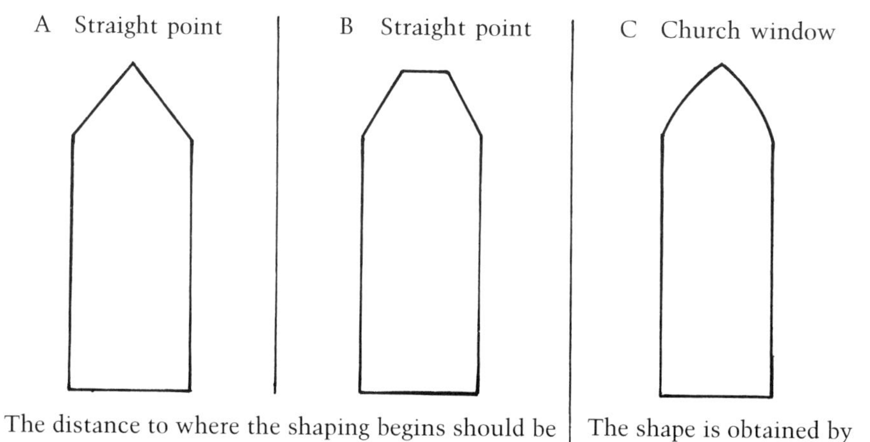

Fig. 3.7 These are the three different shapes in which you can cut the end of a point or strap. B is more traditional for bridle work whereas C is more often used for harness work. Do, however, shape any replacement parts in the style of the item being repaired.

Use a No. 1 or No. 3 edge tool to chamfer the edge on both the grain and flesh sides. When using an edge tool, do not try to dig into the leather, but merely push steadily, holding the tool at an even angle (see Fig. 3.8). With practice you will be able to take the edge off the leather, including curves, in one piece on the grain side. By rounding the edges, it not only looks better-finished, but also helps the point run through a buckle easier, and prevents the sharp edges of the leather from chafing the horse or rider.

To help seal the raw edges, apply a water-based dye to them. The best applicator is a short stick, onto which a felt-type material

Fig. 3.8 By taking off the sharp edge of the leather, you not only make a neater job but also make a more comfortable fit for both horse and rider.

has been tacked. Use a piece of cotton or linen to polish the edges by folding it a couple of times, and pulling the leather through the pad (see Fig. 3.9).

A creasing iron is heated over a flame and when drawn along the edge of the work produces the distinctive dark line running the length of the leather. Considerable practise is needed to master this tool because it needs to be adjusted to the correct measure, by means of a screw, so that the left-hand blade runs on top of the leather, while the other blade runs tight to the leather edge (see Fig. 3.10).

Check that there is a good difference in the height of the blades. If not, then reduce one of the blades with a file, normally the left-

Fig. 3.9 The application of edge dye will both protect and enhance the finish of the leather. A cotton or linen rag is used to polish the edges.

Fig. 3.10 The dark line running along the edge is made by a heated creasing iron. When you first start to use this tool always test the heat on a piece of scrap leather.

hand blade, when looking at the tool in use. A left-handed person may find it useful to file the right-hand blade.

Practise on a piece of scrap leather first because, if the creasing iron is used too cool, it will tend to stick on the surface and not make a mark, but if it is too hot, it simply burns itself deep into the leather.

Point holes

The holes in the point are marked with a compass and, by tradition, are set at the width of the leather for round holes, and at the width plus ⅛ in for oval holes (see Fig. 3.11). From the end of the point, the first hole is marked three widths up. The holes can be made either with a revolving wheel punch or with an oval hole punch and mallet. Before making the holes, first check on a piece of scrap leather the size of the hole required for the buckle tongue in question.

Round holes are traditionally used in bridle work and when using nickel buckles. Oval holes will allow thicker buckle tongues to lie better and are used in heavier work such as stirrup leathers, headcollars, girth straps and driving harness.

Fig. 3.11 All holes, whether they are crew, oval or round, are marked with a compass as shown. Keep the points sharp to help marking.

Point stitching

On a point for a rug or webbing surcingle, the stitching would normally be six or seven to the inch, with approximately 2 in of stitching. Mark these with a pricking iron, just inside the line of the creasing iron, tapping firmly with a mallet. The idea is to mark the stitches by cutting into the top surface of the leather, not drive the pricking iron through the entire thickness of the hide (see Fig. 3.12).

Fig. 3.12 The stitch marks are made with a pricking iron which should only cut the top surface of the leather. The teeth are set at an angle which should coincide with your awl blade when stitching. This way all your stitches will lie at a neat angle in the leather.

The last job before stitching is to carefully skive the end of the point *underneath* the stitch marks and nip off the sharp ends. This will taper the end of the point on the webbing side, so keeping the job as neat as possible (see Fig. 3.13). Because the replacement point is being stitched onto webbing, single needle stitching would be the most appropriate.

The making of a point is the basis for most bridle work because

Fig. 3.13 Skiving off the ends of all buckle returns helps reduce any rough ends that could rub a horse or pony. Do not push the knife forward but rather slide it across the leather at an angle of 45 degrees.

Fig. 3.14 Parts of the bridle.

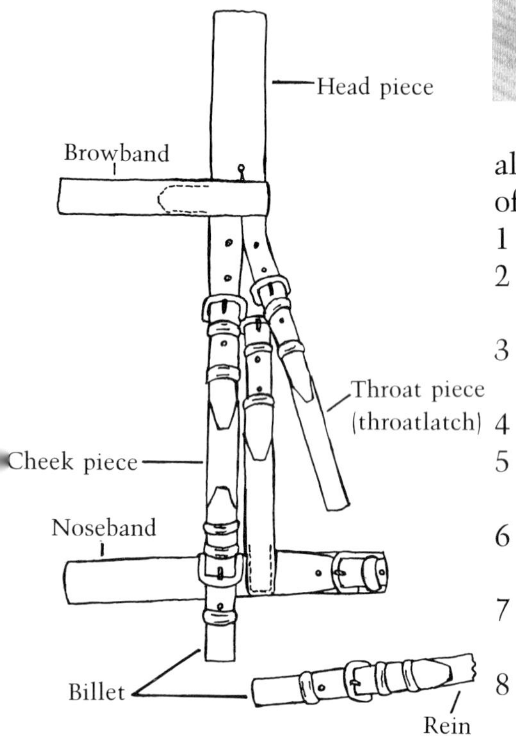

all the processes required are carried out in this job. They consist of:

1 Cutting the strip of leather from the hide.
2 Trimming to the required length and shaping the end if necessary.
3 Edging both the flesh and grain side of the leather for neatness and comfort.
4 Applying edge dye and polishing with a cotton pad.
5 Using the creasing iron to imprint the dark line running along the length of the leather strap.
6 Marking the holes with a compass and punching these out using either a wheel punch or a hand-held punch and mallet.
7 Deciding on the size of stitching (see Fig. 3.15) and marking accordingly.
8 Skiving the underside of any area to be stitched and taking off any sharp corners.

Fig. 3.15 A guide to the most suitable size stitches for each job

PRICKING IRON *5 and 6 stitches to the inch*	PRICKING IRON *7 and 8 stitches to the inch*	PRICKING IRON *9 and 10 stitches to the inch*	PRICKING IRON *11 and 12 stitches to the inch*
Heavy work	Stirrup leathers	Bridle work	Bridle work—best quality
Stable headcollars	Girths	New work and good quality repairs	
Fittings on stable or New Zealand rugs	Good quality headcollars	Fine show-quality pony headcollars	
Surcingle and rug points	Driving-harness repairs		
	Surcingle and rug points		
Use 5-cord or double thickness thread	Use 5-cord or double thickness plaiting thread on the heavier work.	Use 3-cord thread	Use 3-cord thread

For 5-cord, or double thickness 3-cord, use No. 1 or 0 needles. For 3-cord use No. 3 needles.

Other Bridle and Headcollar Repairs

Loops: fixed and running

FIXED LOOPS

One of the most awkward jobs to come to grips with is stitching-in new loops. Nothing looks worse than broken loops with the straps left flapping, especially on the cheek and throat parts of a bridle.

First of all select a thin pliable piece of leather or if this is not available, you will have to skive a piece of leather down to size. It is unlikely that anyone just beginning to do their own repairs will have access to a saddlers' splitting machine. These marvellous, but expensive, machines are bolted onto the work bench and, by adjustment of the hand wheel at the bottom, allow you to thin a piece of leather down to any thickness (see Fig. 1.8). It is perfectly possible to get by without one as far as loop material is concerned, but you will have to learn to carefully skive down the thickness of a piece of leather by hand.

This is best done on a piece of leather longer and wider than the actual size of the loop to be used. Hold the knife at a low angle and draw it across the leather at an angle of approximately 45 degrees. Do not try to push the knife straight ahead because you will only succeed in cutting completely through the thickness of the leather. Once you have a piece of loop material of similar thickness to the one being replaced, measure the correct width with a compass and trim to this size. The length of the loop must allow for the thickness of the leather point which is to go through it, plus the extra amount stitched in to the strap.

Do not attempt to use the edge tool on the loop unless it is a heavy loop from, say, a piece of driving harness, merely dye the edge and use the creasing iron set to a very fine measure. There are two methods of stitching loops. First I will explain the correct and normally practised way used by saddlers, and second, a simpler, but nonetheless effective, way you may like to try.

Try not to start with a replacement loop on a riding bridle throat piece. Even with years of practise they can still occasionally go wrong. Instead try something of about one inch in width.

Fig. 3.16 Before starting to stitch, mark the middle of the width of the leather on the loop. This should help you to stitch the correct amount of leather on each side.

Hold the work in the clamps with the buckle away from you and mark the middle of the width of leather on the loop (see Fig. 3.16). Insert the loop up to the mark and stitch with either one or two needles straight through the loop, and, using the same method used for stirrup leathers, continue stitching along the first side of the loop, cross the thread over, and stitch along the second side of the leather to where the loop tucks in on the other side of the strap (see Fig. 3.17).

Fig. 3.17 With the work held firmly in the jaws of the stitching clamps, start stitching towards you. In this picture the loop has been held in by the first three stitches.

Now for the fiddly bit. If there are no holes in the strap where the loop is to fit, make holes first by pushing the awl through each stitch that is marked but, without drawing the thread through. This will give you a guide hole to follow. Fold the loop around the strap and position it between the thicknesses of leather.

You should have cut the loop with sufficient extra length, so when the two ends meet inside the strap being repaired, you should still be able to stitch underneath the loop by making

Fig. 3.18a When stitching the return of the loop, make the holes with the awl before inserting the loop. The awl can then be gently pushed through again at a slight angle so catching the loop, but do not use the awl as a lever or it will break.

Fig. 3.18b Having made one stitch to hold the loop in place, pass the right hand needle and thread through the loop and continue stitching on the near side of the loop.

a stitch or two on the far side of the loop, then gently pushing the loop forward with your fingers and stitching on the near side of the loop (see Figs. 3.18a and b).

When finished, the loop needs to be blocked-out to make it square. This allows the strap to pass through with ease. Slightly dampen the loop before pushing the correct size loop stick through and lightly tap it to shape with a hammer (see Fig. 3.19). Take care to only hit the loop against a flat surface of the loop stick, otherwise a blow against an edge could cause the loop to split.

Using a cold creasing iron on the loop tops not only provides an attractive finish, but also helps the loop to keep its shape (see Fig. 3.20).

A slightly more simple method of stitching fixed loops is to first make it in to a complete loop by stitching it together end to end (see Fig. 3.21). This will ensure the loop is secured together, but, when stitching, care must be taken to place the join in the middle of the buckle return where it will not be seen.

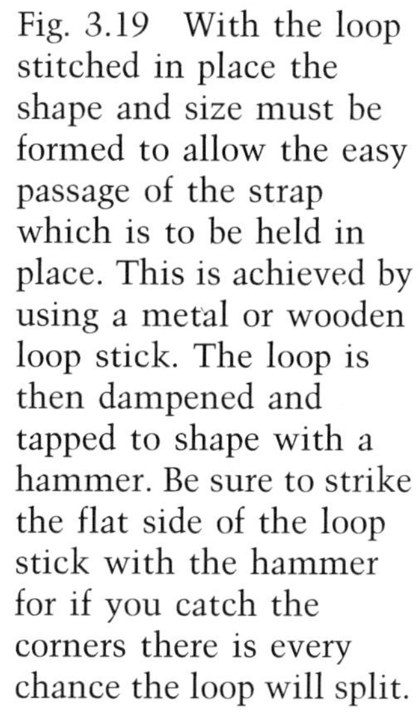

Fig. 3.19 With the loop stitched in place the shape and size must be formed to allow the easy passage of the strap which is to be held in place. This is achieved by using a metal or wooden loop stick. The loop is then dampened and tapped to shape with a hammer. Be sure to strike the flat side of the loop stick with the hammer for if you catch the corners there is every chance the loop will split.

Fig. 3.20 A cold creasing iron run over the tops of the loops will make the attractive dark lines in addition to helping the loop to keep its shape.

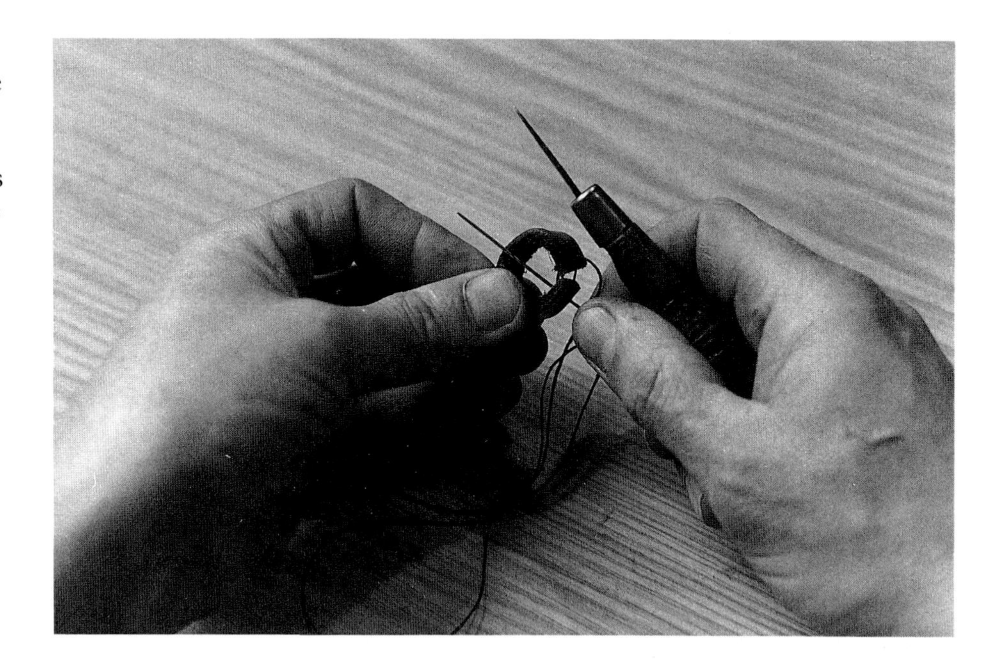

Fig. 3.21 Making a loop up in this fashion, before stitching it into place, is not general saddlers' workshop practice, but is a simpler method for the novice.

Important Note. When stitching loops, more than at any other time, great care must be taken to push the awl through the leather with a straight movement, i.e. as near as possible an angle of 90 degrees to the work being stitched. **Do not** try and lever the loop into position, because this will often result in a broken awl blade.

It is far better to cut the loop slightly larger than smaller, to facilitate ease of stitching, as well as giving ample room for the leather point to pass through.

RUNNING LOOPS

These are the loops that move or run along the strapping and keep the end of a point in place.

The simplified method of stitching a loop just mentioned could be employed here, but in normal workshop practice you would overlap the ends of the loop – rather than placing the ends of the loop edge to edge – and use three or four stitches to hold it together.

Prepare the loop material as for a fixed loop but allow an extra quarter inch or so to accommodate the extra stitches. Mark the

Fig. 3.22 Stitching running loops can be one of the most fiddly jobs you are likely to attempt. Stitching the loops held by a pair of upturned compasses in the stitching clamps is the best method I have found. The loop overlaps by three or four stitches and is stitched together with one needle.

stitches on the end of the loop using a 7 or 9 to the inch pricking iron.

Over the years I have tried many different ways of holding running loops while stitching and have found the easiest way is to use a pair of compasses held upside down in the stitching clamps. Fold the loop around so that the stitching marks line up with the overlap at the other end of the loop and stitch with one needle (see Fig. 3.22).

Headcollars

CHEEKS, JOWLS AND BACKSTAYS

These are the parts that most commonly need repairs (see Fig. 3.23). A ring at the back of the nose and jowl pieces is far stronger than a square and will also put a little less stress on the leather of the jowl ends (see Fig. 3.25).

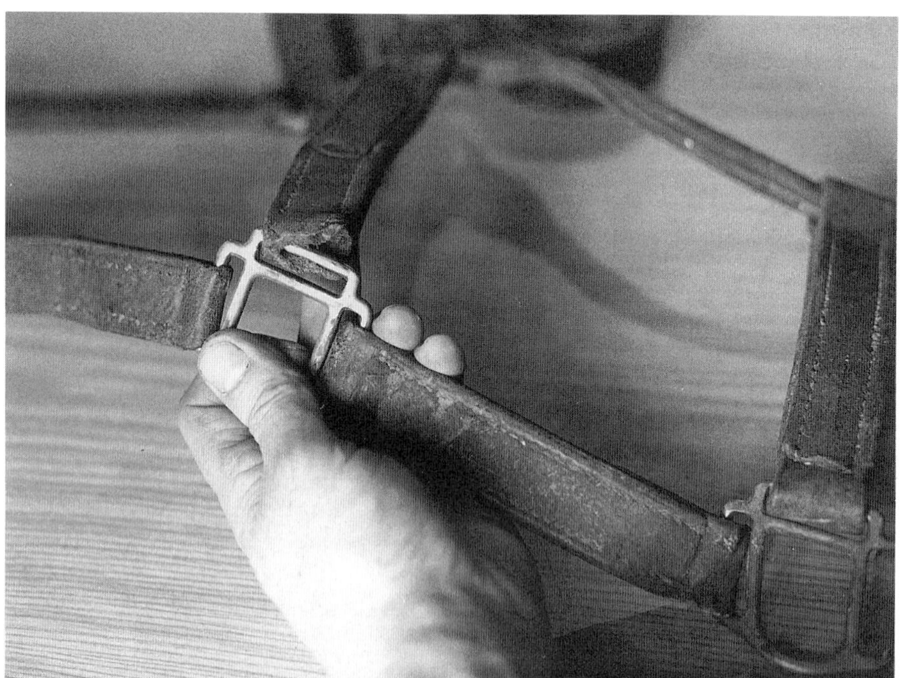

Fig. 3.23 The end of the cheek piece can clearly be seen to be in immediate need of repair. In this case the metal square has held but the leather needs replacing.

Fig. 3.24 Parts of the headcollar.

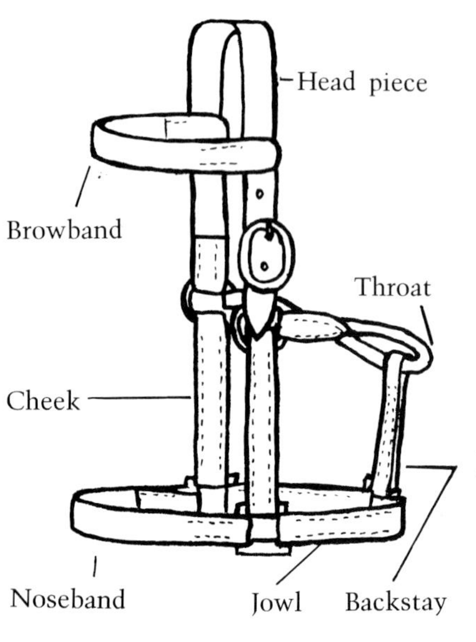

There are two ways to effect repairs to these pieces.

1 Use an insert. Cut a piece of leather approximately 5 in long × the width of the broken piece. Next cut the end off

Fig. 3.25 At the back of a headcollar you will find either a metal ring or a square. A ring is generally stronger because the stress put upon it is distributed around its circumference and not on any one angle, as with the square. A ring can distort but is less likely to break in use.

the worn piece and split the stitches to make room for the new piece when it is folded double (see Fig. 3.26). Pull out all the old stitches you have just cut, insert the new piece and stitch with two needles.

Fig. 3.26 The broken jowl of the headcollar needs to have a new piece of leather stitched into it at one end. The new piece has had both ends skived down thinly so it will form a neat splice with the old leather.

If the cheek or jowl piece is badly worn, then the insert could be stitched on the outside of the main piece when it looks more like a patch.

2 For a far better repair, completely replace the broken piece. Carefully cut the worn piece off, splitting all the stitches; this way you will have a pattern from which a completely new cheek, jowl or backstay can be cut out. Prepare the leather and mark the stitches as closely matched to the rest of the headcollar as you can, and using the heaviest thread you have, double stitch the leather.

THROAT PIECES

Occasionally, the throat piece will need repair. If this is a flat throat, i.e. a solid piece of leather with stitching only at each end, it is easy enough to simply splice (overlap the broken ends of the throat piece by two inches and stitch together) or make a new end. If, however, it is a rolled throat piece with stitching running along its entire length, then, unless it only needs a new end, this is going to require a very specialist repair. A simple way out, and one which is perfectly acceptable is to replace the worn rolled throat piece with a flat piece of leather of the same length (see Fig. 3.27).

Fig. 3.27 A broken, rolled throat piece like this can either be shortened by a couple of inches with part of the rolled work flattened and inserted into the return, or a replacement throat piece will have to be fitted to the headcollar. I feel it is quite acceptable to replace a worn rolled throat piece with a flat throat piece.

HEAD PIECES

With constant use the holes on the head piece can start to split or tear out completely. A small tear can sometimes be stopped by putting a couple of stitches below the hole to hold the split together. If, however, the head piece is in need of total replacement make this up using the thickest leather you have to hand that is of the same width as the original piece. An old stirrup leather could well fit the bill here, so, as I have mentioned before, never throw away any pieces of tack if the leather is in good order.

Leather headcollars never seem to get as regular a clean and oil as, say, a bridle, but often have more stress put upon them. Once the leather has been neglected and gone hard they can be very difficult to stitch and repair. So give them a bit more attention and most of these repairs can be avoided.

4
THE SADDLE

The largest and most expensive piece of equipment you will own, barring a quality set of driving harness, is the saddle. A good saddle that is well looked after, regularly cleaned and oiled, and repaired properly when needed, can outlast an owner's riding lifetime. In addition, once you are used to your own favourite saddle, no other saddle is quite the same. It makes more than just economic sense, therefore, to take as much care of your saddle as possible.

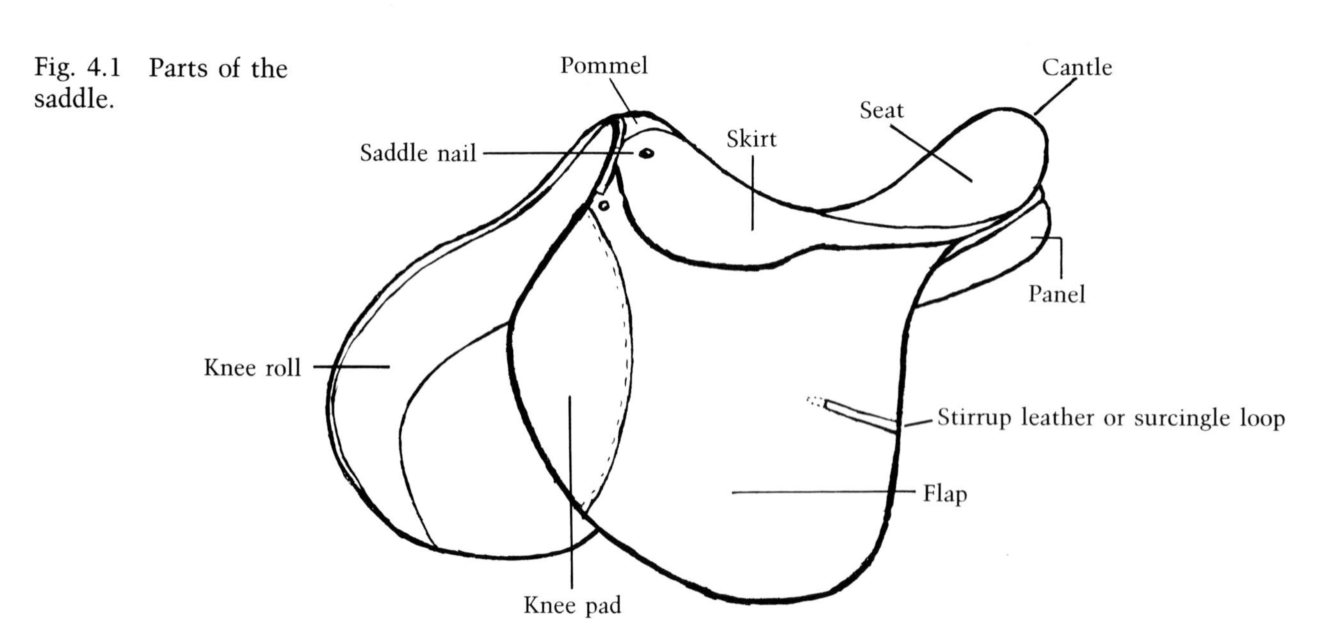

Fig. 4.1 Parts of the saddle.

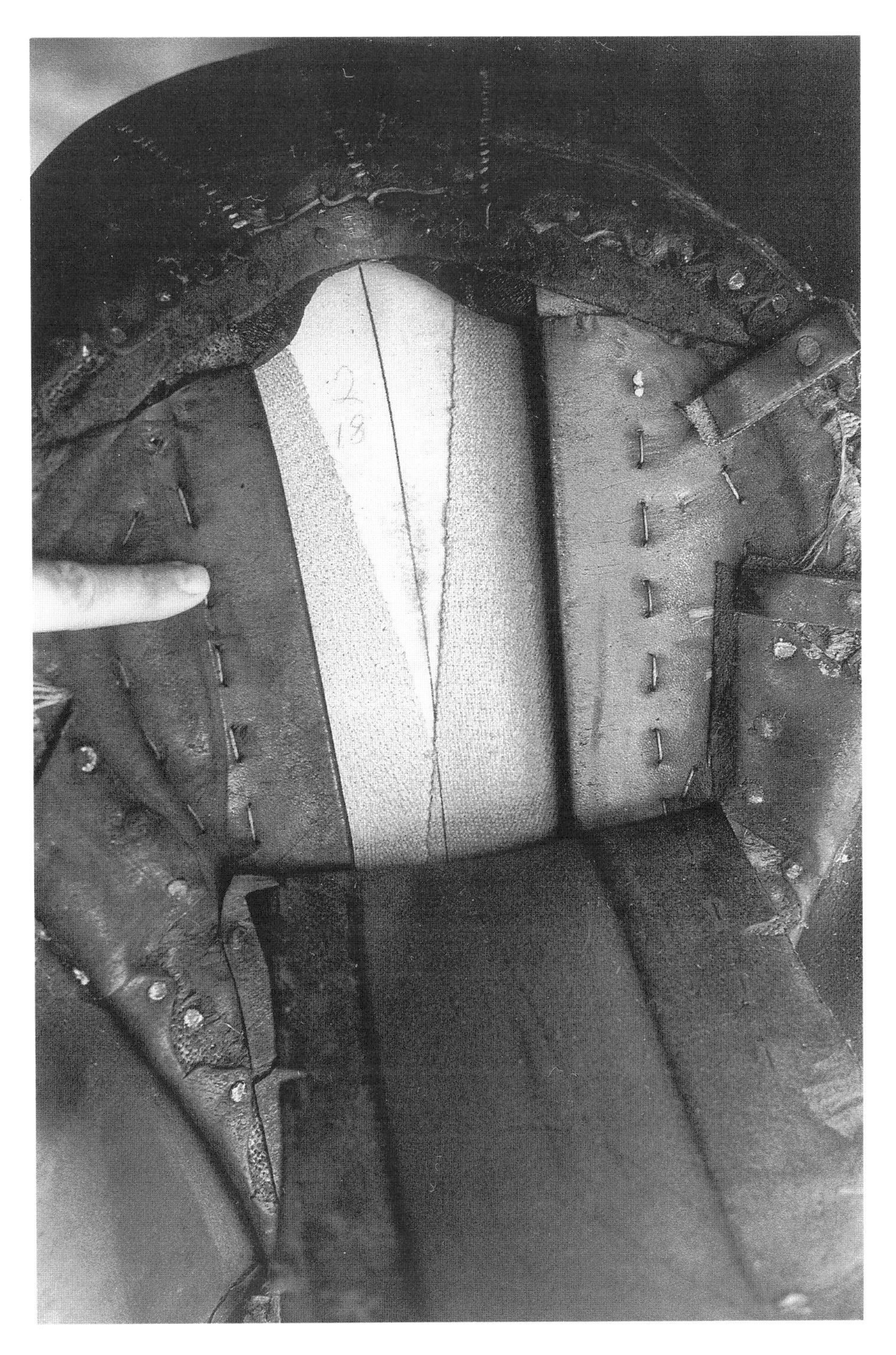

Fig. 4.2 This picture is a good view of the underneath of a saddle seat with the panel removed showing the flat springs neatly covered in leather and pop stitched. The numbers in the middle indicate the manufacturers batch number, and the long seat webs are clearly seen each side forming a V as they meet towards the narrower part of the seat. All this is then covered by a thin piece of hide and the panel laced into place.

Remember that, with a saddle, the outer surfaces are going to be in contact with either you or your horse, so keep this in mind all the time you are working on a repair and strive for not only neatness, but smoothness.

Girth Straps

No matter how well a saddle is cleaned and oiled, sooner or later the girth straps are going to need restitching or replacing.

Certain close-contact and dressage saddles have easier access to the stitching owing to their extended girth webs onto which the straps are stitched. With nearly every other type of saddle you will have to drop or take down, the front of the panel first by carefully cutting the stitches just in front of the pommel.

Start where the flap is first joined to the panel, not at the centre of the head, and gently prise them apart as you nick each stitch (see Fig. 4.3). Then, gently but firmly pull the tree points out of the pockets of the panel, so opening up the front of the saddle to work on.

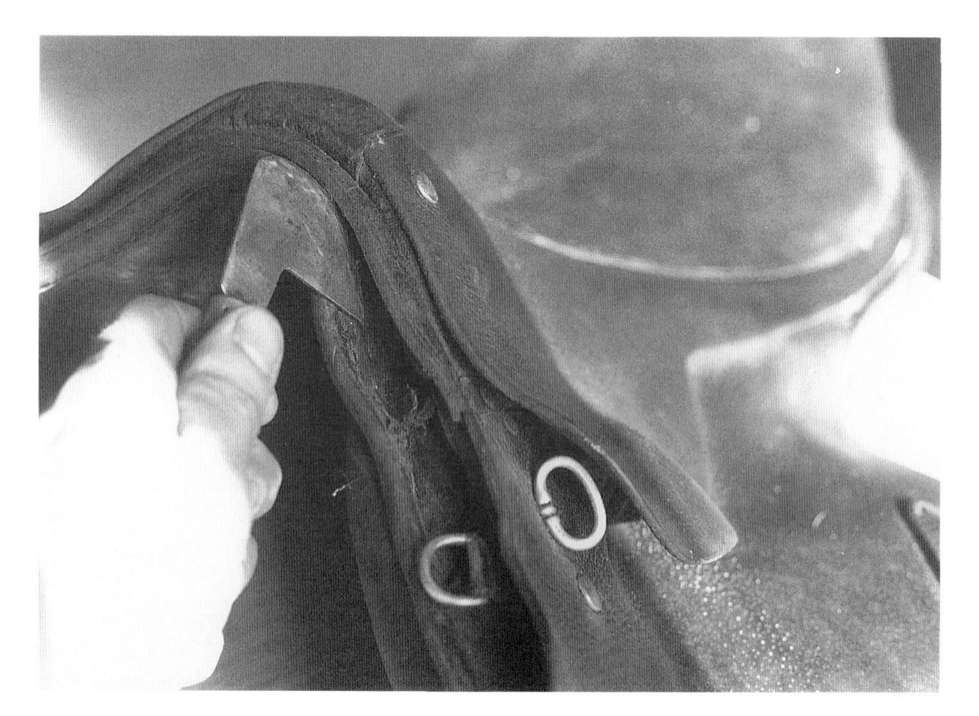

Fig. 4.3 The front of the panel is attached to the top of the saddle by approximately six to eight pop stitches. In this picture these large stitches are being cut with a head or half-round knife.

I have come across only one other saddle on which you can work on the girth straps without first cutting the stitches holding the panel to the saddle, and that is the Turf and Travel Cavalier. It is a well-made ultra-lightweight saddle that reduces the weight and thickness between horse and rider by combining the flaps with the panel. The girth straps are stitched to longer webs than normal and passed between the flaps-cum-panels at the bottom end. Unlike the dressage saddle you cannot gain direct access to the stitching on the straps, but by slipping your fingers underneath the flocked part of the panel along the channel you can, perhaps with some difficulty, just reach the webs and pull them through. There have been several copies of this saddle over the years, but they should all have the same access to the girth straps.

When you have gained access to the stitching cut the girth straps off and, using an old one as a pattern, cut the new ones to size. The nearside straps will normally wear first owing to the horse being girthed up on that side. Having pointed one end of the strap, edge it with the largest edge tool you have, then dye and polish the strap. With the exception of chrome leather and buffalo hide, which will not normally take an edge tool, girth straps are made up with the flesh or underside as the top surface. Mark and punch the holes using only one of the old girth straps as a guide for all the new ones, because the old straps will have stretched at a different rate causing uneven holes.

Girth straps are one of the very few pieces of saddlery that have the stitching holes punched completely through before stitching commences. This can be achieved with either a large pricking iron with round pointed teeth or, just as easily, with an ordinary revolving wheel punch; the holes being ¼ in apart and marked with a compass. You will need approximately six stitches each side not the two or three used by some saddlery factories (see Fig. 4.4).

With certain cheaply made saddles, you will find that the girth straps have been nailed directly onto the wooden tree. This is an appalling practice when making *new* saddles but may be the only way to effect a *repair*. The stress and pull of the girth is transmitted directly to the nails in the tree and not over the entire waist of the saddle, as would be the case if they were stitched

Fig. 4.4 After cutting and preparing a replacement girth strap, the stitch holes are then made with either a large type of pricking iron or, as in this case, with a wheel punch. This is one of the few times when the stitch holes are preformed, and, because a heavy thread is to be used, are well spaced apart. Note that there will be six stitches each side, in addition I always stitch over the edge on the first stitch each end.

to a cross web. On occasion, I have had to nail a replacement girth strap onto the tree if the tree webs had rotted, but I would always try to fit a new web, be it unorthodox, if at all possible. I say unorthodox because, when a saddle is made, the girth webs pass over the top of the waist of the saddle tree before any leather is attached. The repair is made by fitting a replacement web underneath the original saddle webs, and gluing and tacking them into place. In this way the stress is taken in part on the existing saddle webs, not just on the tree (see Figs. 4.5a, b and c).

So, wherever possible stitch, do not nail, replacement girth straps. Use double thickness 5-cord or three or four strands of plaiting thread on one needle. You will not be able to hold the work in your clamps – except in the case of the Cavalier saddle – but must hold the strap to the web with one hand and back stitch with the other (see Fig. 4.6). Tap the stitches down with a hammer when finished to reduce friction when in use.

Replace the saddle tree points into their pockets and reposition the front of the pommel in its correct place with the head of the saddle. That is, when viewed from the front, the start of each panel should be equidistant from the centre vertical of the saddle

Fig. 4.5a This is an example of a badly fitted girth strap. The web onto which it is stitched is only nailed to the tree and does not pass over the lateral, or the long, webs of the saddle as does the wider girth web. All the stress on the narrow web is passed directly onto the saddle tree.

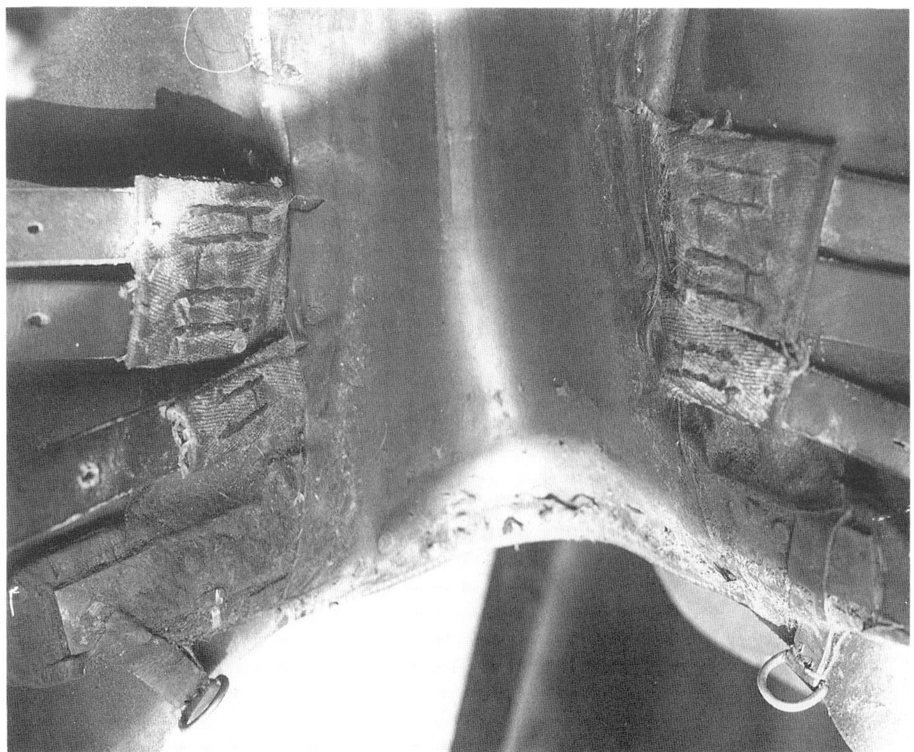

Fig. 4.5b This picture clearly shows the two cross webs on a saddle. The wider one carries two girth straps, and the narrower web, one strap. When studying this picture, the advantage of always using the front girth strap on the narrower web plus either of the other two when girthing up can be appreciated; if one of the webs should break then the saddle will still be held on by the other.

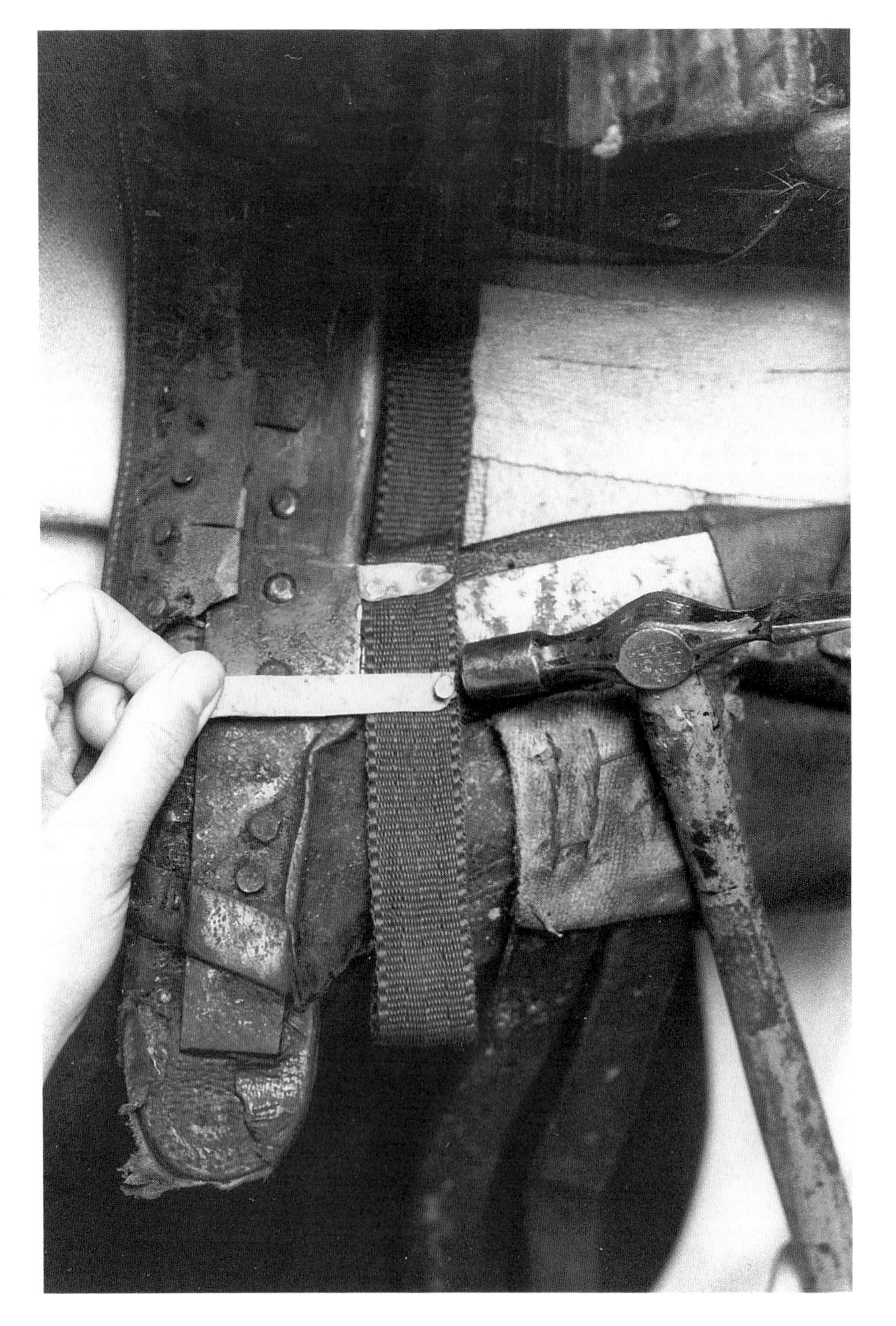

Fig. 4.5c A safer job is achieved by fitting a replacement front web that passes underneath the long webs of the saddle. This double thickness length of web is glued onto the long webs and tacked into place using a strip of leather between the web and the tack head. Time and damp will always cause tacks to rust a little, and the narrow strip of leather helps protect the webs from rust eating into them. With this method, if the tacks should pull out, the girth strap will still be held to the saddle by virtue of a one piece web.

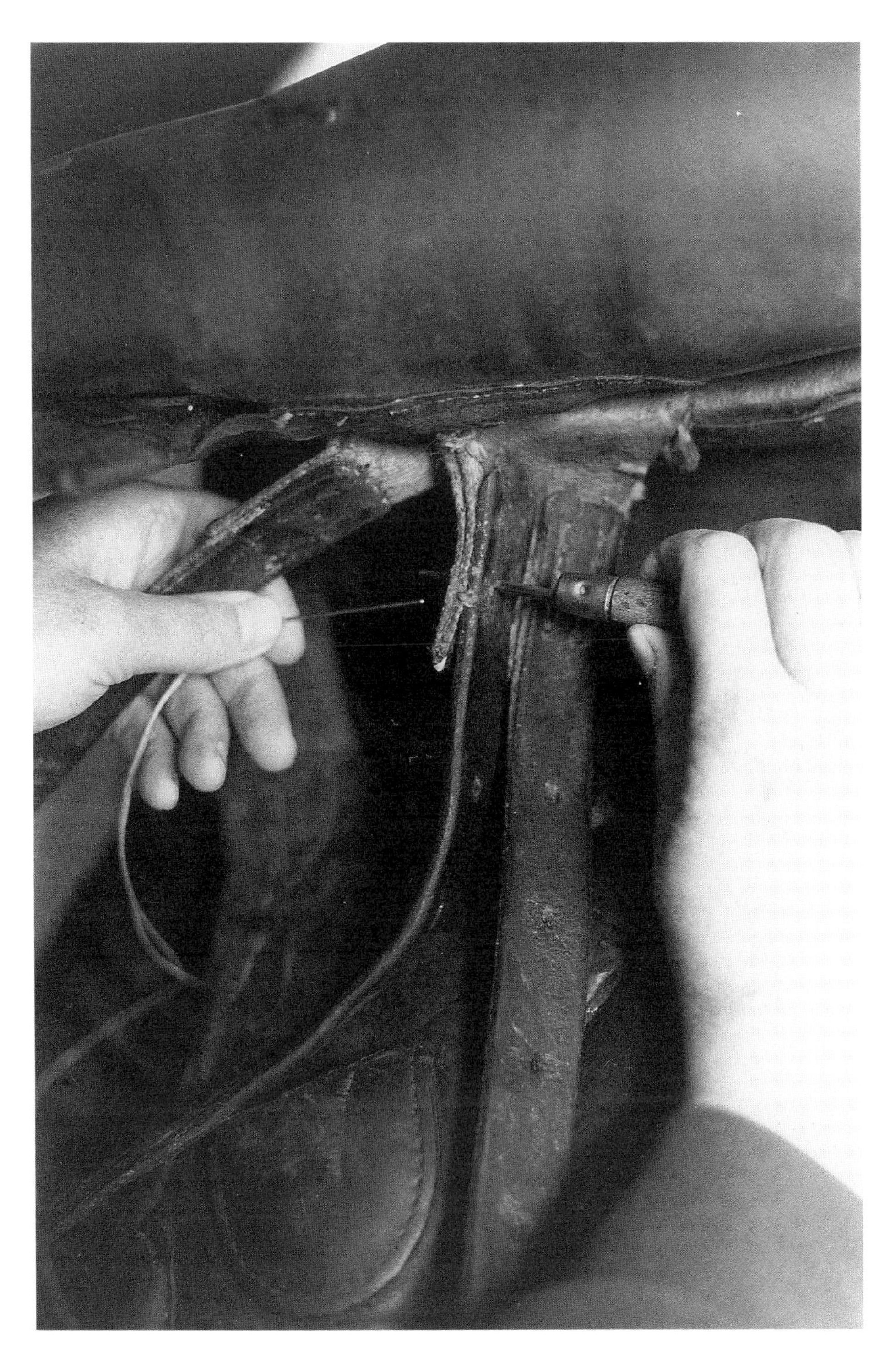

Fig. 4.6 Stitch the girth straps, either new or existing, with a heavy thread, double thickness 5-cord for example, and wax well. After stitching it reduces wear if the stitches are tapped flat with a hammer.

(see Figs. 4.7 and 4.8). You will see that this is easier than it sounds and will find that the stitch holes will line up opposite each other. With certain saddles you may find that the panel has been nailed in place rather than stitched. If this is the case then remove the nails with a nail claw.

When the required repairs have been finished and the panel is properly aligned, it needs to be restitched or pulled up.

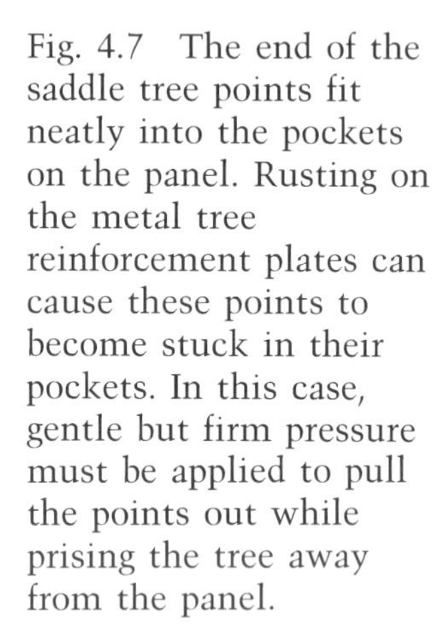

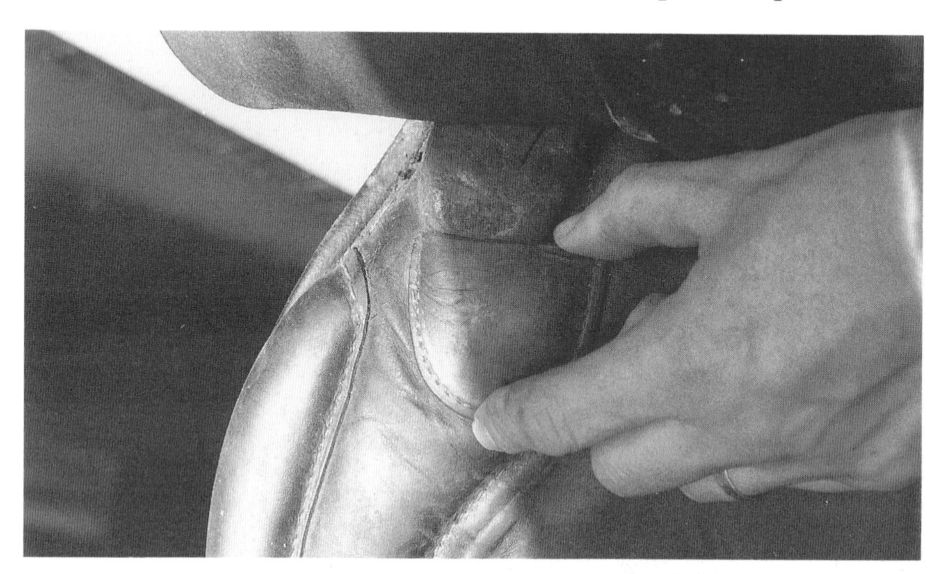

Fig. 4.7 The end of the saddle tree points fit neatly into the pockets on the panel. Rusting on the metal tree reinforcement plates can cause these points to become stuck in their pockets. In this case, gentle but firm pressure must be applied to pull the points out while prising the tree away from the panel.

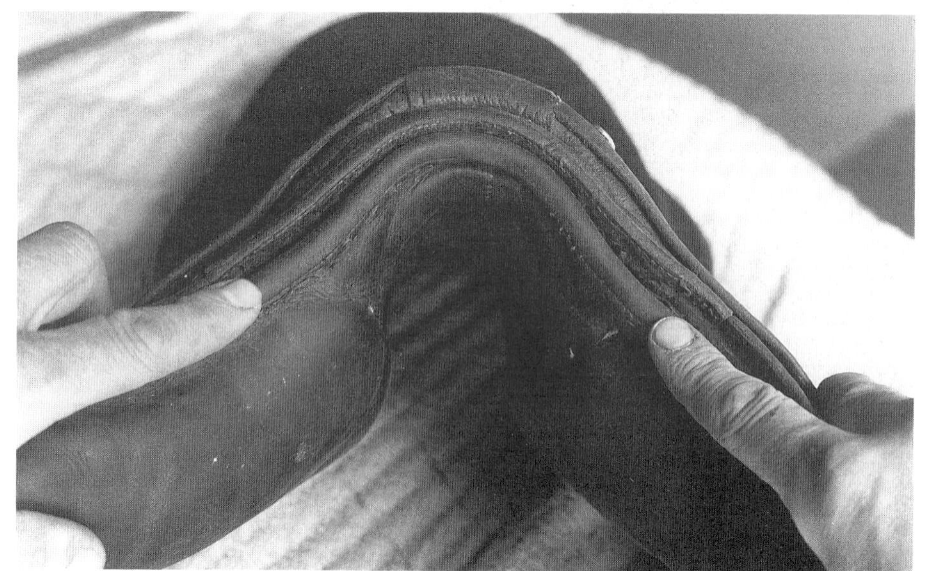

Fig. 4.8 Once the points of the tree are in their pockets, make sure that the panel is level before pulling up. This can best be judged by noting the distance from the start of the stuffing in relation to the saddle nails in the skirts or the seam of the seat.

Restitching the Saddle Front or Head

Wax a piece of double thickness 5-cord thread and thread it onto one needle. Position the needle in the middle of the thread and tie the two ends of the thread together in a large knot.

Start at one end by drawing your needle from the inside to the outside of the panel so that the knot at the end of the thread will be hidden from sight when in use. Make your stitch holes, using the largest awl you have, through the panel and fore piece of the saddle, pop stitching as you go (see Fig. 4.9). (The fore piece is the narrow strip of leather about 10 in long that joins the two saddle flaps, and runs underneath the pommel.) Finish up by taking the needle through to the inside of the panel and tying a knot.

Except in rare cases where the fore piece of the saddle has rotted or there was not one there to begin with, it is totally wrong to nail the front of the panel to the saddle tree. The nails can work loose and fall out, with dire consequences to the poor horse.

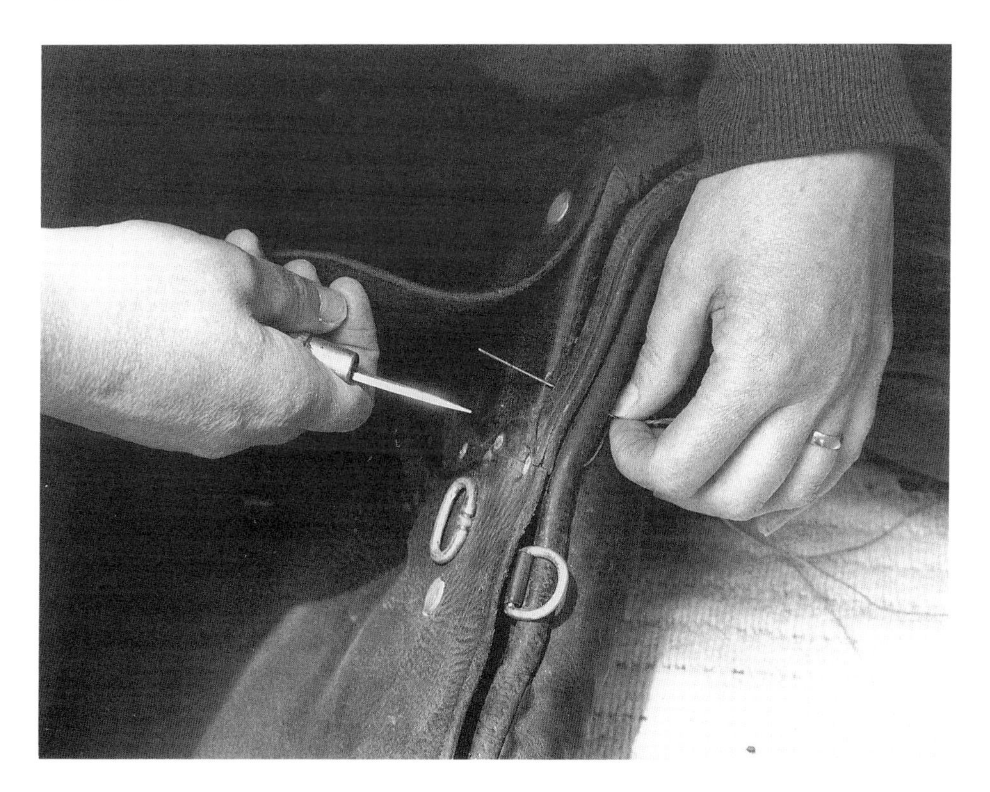

Fig. 4.9 When pulling up the front of a saddle use one needle and a heavy thread. Make the holes between the saddle seat and the fore piece and straight through the panel. Discount the sloppy practice of some saddlers of nailing the front of the panel directly into place. It is only marginally quicker, and could cause untold damage to a horse if a nail should fall out while the saddle is in use.

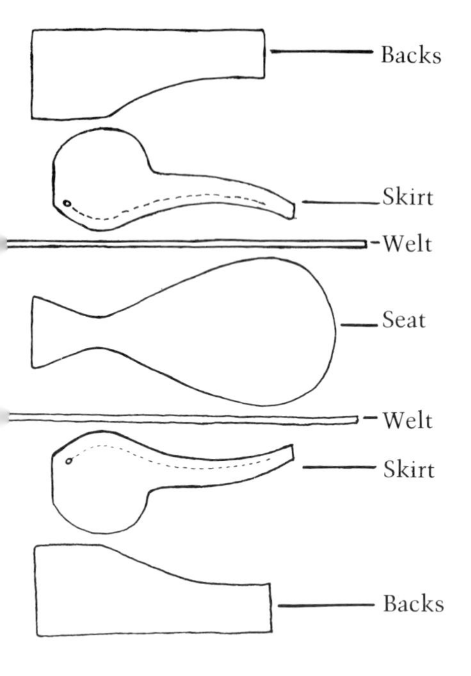

Fig. 4.10 Exploded view of the saddle seat.

Repairing the Saddle Seat

The saddle seat is the thin covering of leather on the top of the saddle. I have often been asked to *just* replace a saddle seat because either the pommel or cantle needs patching, or else a couple of inches or so of the seat has split along the welt. The look of horror on the face of the customer when I quote for *just* replacing a saddle seat in the traditional way says it all.

I have to explain that to do the job properly, the saddle must be stripped down to the tree, and a piece of pigskin twice the size of the actual seat has to be soaked, stretched over the saddle, and cut to shape when dry. The backs, which are stitched onto the skirts normally need replacing and new welting fitted, if, that is, the skirts will take the stress of restraining. Considerable stress is put on the skirts when they are strained over the saddle seat with a pair of web pliers, and ageing leather may not be able to cope.

So, unless a saddle is relatively new or has suffered accidental damage – like the new dressage saddle a customer left on the stable door, only for her promising youngster to tear half the seat out with his teeth – it is not, in fact, financially worth repairing in the traditional manner. There are, however, two other ways in which a seat can be repaired.

Overlining

When overlining, the original surface is left in place and the seat, or panel, is completely re-covered. It is not the sort of job to undertake when just learning saddlery skills, but once you are used to the feel of an awl and needles and thread in your hands, and have mastered the art of simple stitching, there is no reason why a job like this should not be attempted. It may appear as if everything is held together with glue, but by gluing a patch, or lining, in place and letting it dry, it will remain in the correct place, thus allowing you to concentrate on the stitching.

To overline the seat, a large piece of thin hide such as panel hide is glued over the entire surface of the seat and two skirts. The excess is trimmed off the edge of the skirts and the hide is then stitched around the edge to avoid separation when the saddle is in use. Excess hide is then folded over the cantle and laced with a curved needle to the back of the panel. At the front, the lining hide is stitched through the fore piece and pommel before being trimmed to size (see Fig. 4.11).

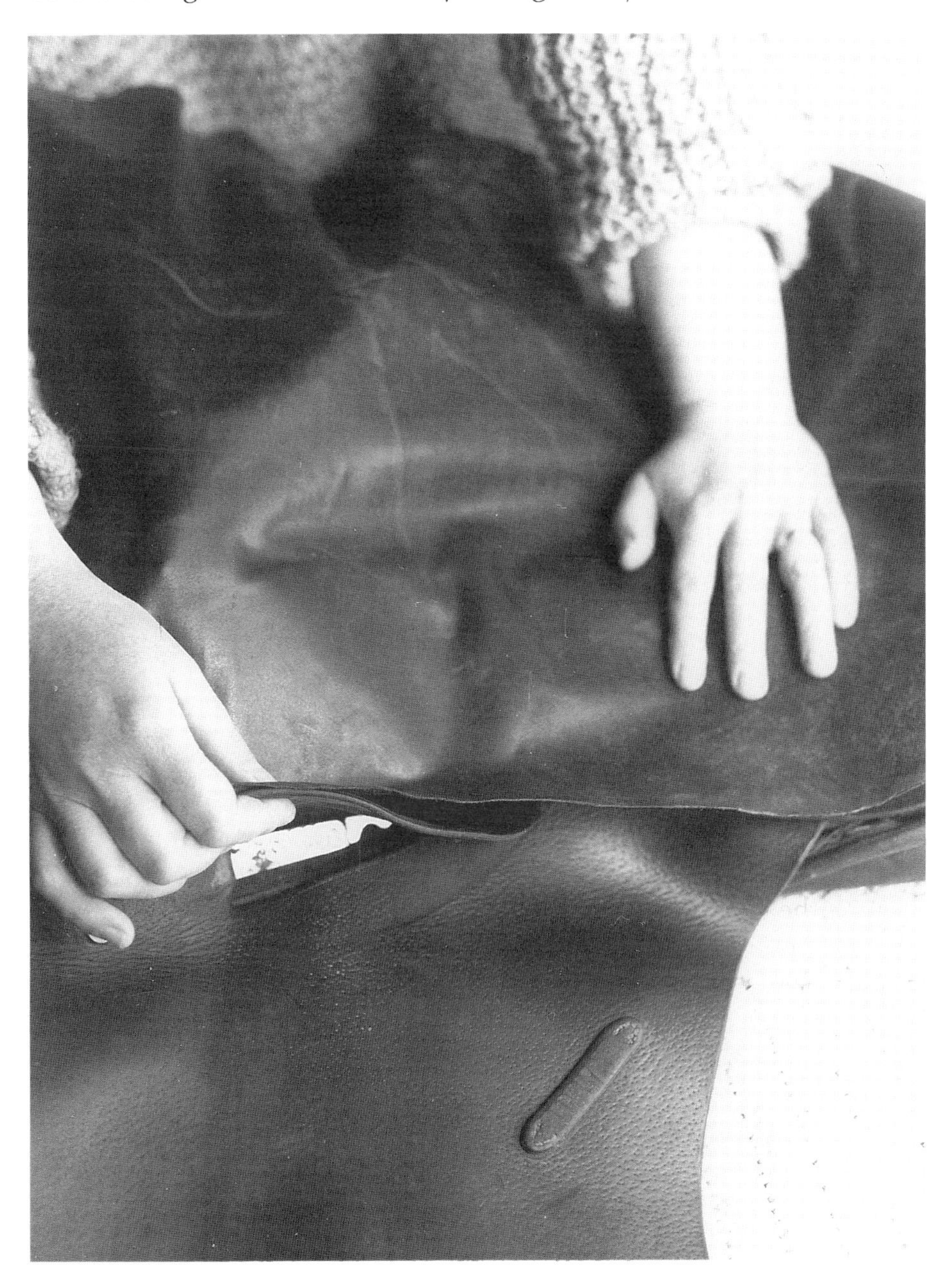

Fig. 4.11 With old saddles, this is sometimes the only way to re-cover a badly worn seat. A thin hide is used to completely cover the seat and skirts each side. Although not an ideal repair, it can last well and prolong the life of a much loved old saddle.

On modern deep-seated saddles this type of repair can look unsightly, so, if the skirts are in good condition, a neat patch to either the pommel or cantle is to be preferred. An old English flat-seated hunting saddle, however, could well benefit from this treatment especially if there are any holes worn in the skirts caused by the stirrup bars or buckles.

Patching

PATCHING THE POMMEL

The first job is to drop or take down the front of the panel and remove the fore piece (see Fig. 4.12). A piece of thin hide is then cut to a size that will cover the damaged area. The patch should have plenty of excess leather in front of the saddle, which will be

Fig. 4.12 To remove the fore piece, use a nail claw close to the saddle tree and tap firmly. Replacing the fore piece is made easier by being able to re-use the original nails and holes.

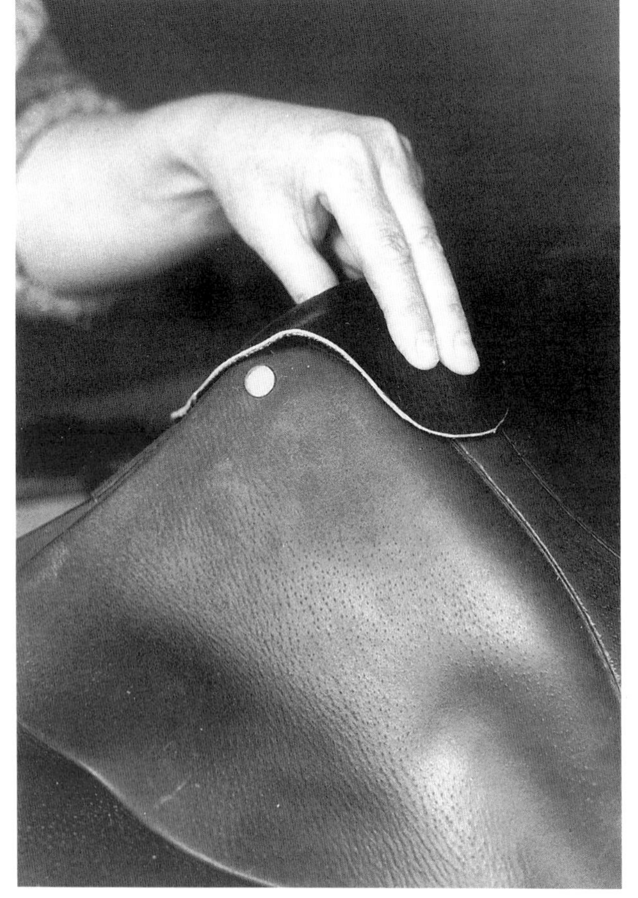

Fig. 4.13 If only the front of the seat has worn or been damaged then a neat patch can offer a reasonable repair.

pulled underneath the head, and at the other end there should be just enough leather to pass the wood and metal tree arch (see Fig. 4.13).

If the damaged area extends to the top of the skirts, cut the patch wide enough to extend just past the saddle nails on the skirts. This larger patch will necessitate the removal of the large-headed saddle nails. You will find that these heavy nickel- or brass-headed nails have been turned over after passing through the tree; these will need straightening with a nail claw before tapping out. Do not try to remove the saddle nails without first straightening their shanks from inside the panel where they have been flattened against the saddle tree.

Next, cover both the patch and the area to be repaired with a good glue and position the patch in place. If the saddle nails have been removed, replace them once the patch is glued on in order to keep it in place. I normally then leave the patch to dry for at least a couple of hours or so, rather than have the patch slide around while I am trying to work on it.

To stop the part of the patch nearest the rider lifting up when in use, you will need to stitch about two to three inches along the edge, stitching through the patch, saddle seat and tree webs. Although this sounds heavy going, it is quite easily achieved, but care must be taken not to snap the awl blade by pushing it into the metal of the saddle-tree arch. You will be able to feel and see where the arch starts and finishes. It is the small part of the patch that extends beyond this that needs stitching if the repair is to be made permanent (see Fig. 4.14).

The stitches are best marked with a compass approximately ¼ in apart once the patch is glued in place and the position of the tree arch is ascertained. Stitch using a single needle with 3-cord thread.

Once the patch is secured this far, the excess leather in front can be dampened with water, folded over the front of the saddle and fixed into place with a hammer and tacks. Any surplus leather can then be trimmed off. Refit the fore piece with tacks and pull up the pommel as previously described.

When properly done, this repair can look very tidy and will last the life of the saddle. Once oiled to match the saddle, it is

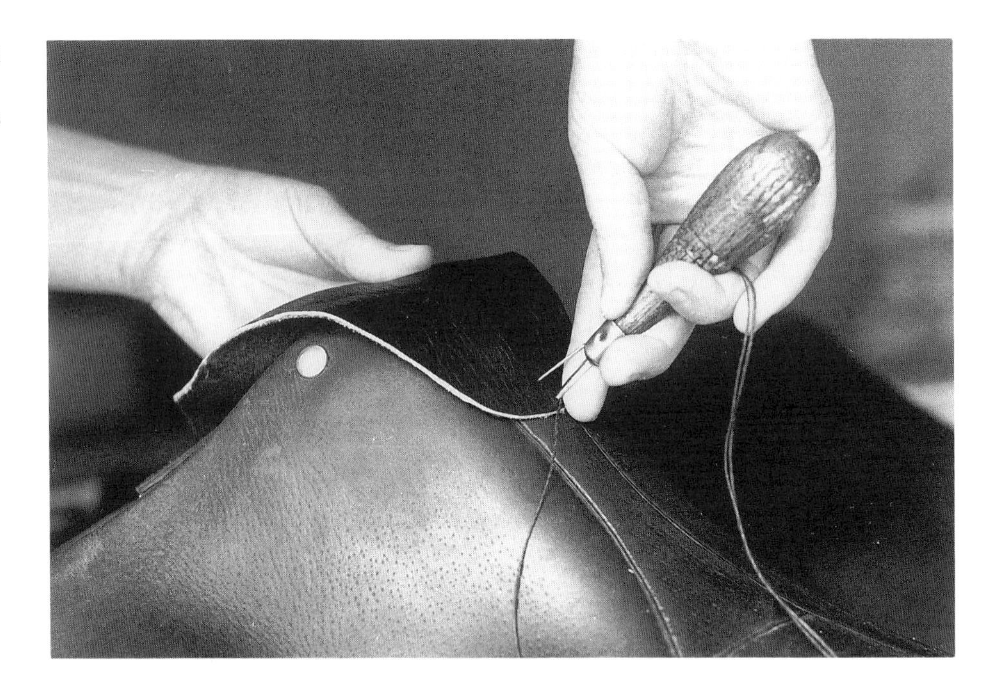

Fig. 4.14 With the patch glued in place, stitch through the saddle seat as far as you can until you reach the tree arch. The patch would normally have the edges stained before stitching, they have been left white in this picture for the sake of clarity.

very difficult to see the repair.

An easier and quicker, but not nearly so permanent, way to patch the pommel is to simply cut a piece of thin leather to shape and glue it directly to the seat without any form of stitching or tacking. This will hold for a while, but sooner or later it will start to peel of and need regluing.

Note: if the saddle seat has been oiled regularly and a patch is to be glued onto it, first mark out the size of the patch on the oiled leather, and rough the surface with a knife edge or course glass paper. This will give a key to the surface which will help the glue to grip.

PATCHING THE CANTLE

The most common problem with the cantle is that the top edge splits revealing the foam or surge and flock seat padding, or even the tree itself. If it is just a split, that is, no leather is actually missing, then a good repair can be effected by cutting a narrow strip of very thin leather just longer than the split, this is then carefully inserted inside the split part and folded over the edge of the cantle. A little glue underneath the patch will keep it in place. Allow the glue to dry for a while and then feed glue underneath the saddle seat using a thin piece of scrap leather. Press tightly against the patch until the glue begins to get tacky. Leave to dry. Then take two needles and a length of thread, of the colour nearest to that of the saddle seat, and very carefully stitch along the edge of the split, through the patch and onto the saddle seat at the back of the cantle. (see Fig. 4.15). Do not pull the stitches too tight, just enough to draw all the leather together.

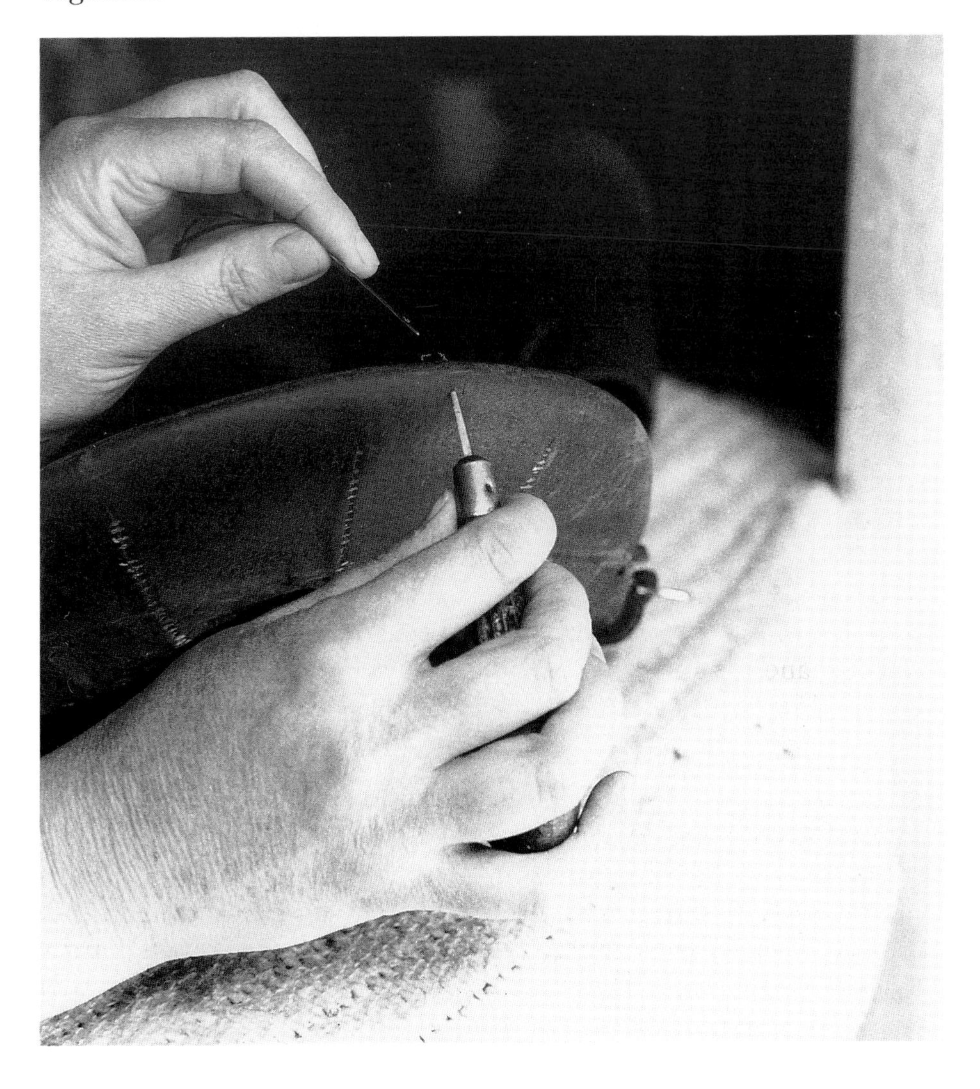

Fig. 4.15 Splits on the cantle part of a saddle can be repaired by carefully inserting a thin piece of skin inside the split, gluing, and then stitching through the seat patch and even through the top half inch of the wooden saddle tree. Do not pull the stitches too tight, just firm enough to hold everything together.

When the glue is completely dry, melt a little beeswax over a flame and rub into the stitching and the split (see Fig. 4.16). Wait a few minutes for the beeswax to dry and then rub lightly over with a bone or other smooth tool such as a wooden handle. All excess beeswax will come off and, with care and a little practise, will completely fill the stitch holes making a near invisible repair. If however you rub off too much beeswax then simply melt a little more.

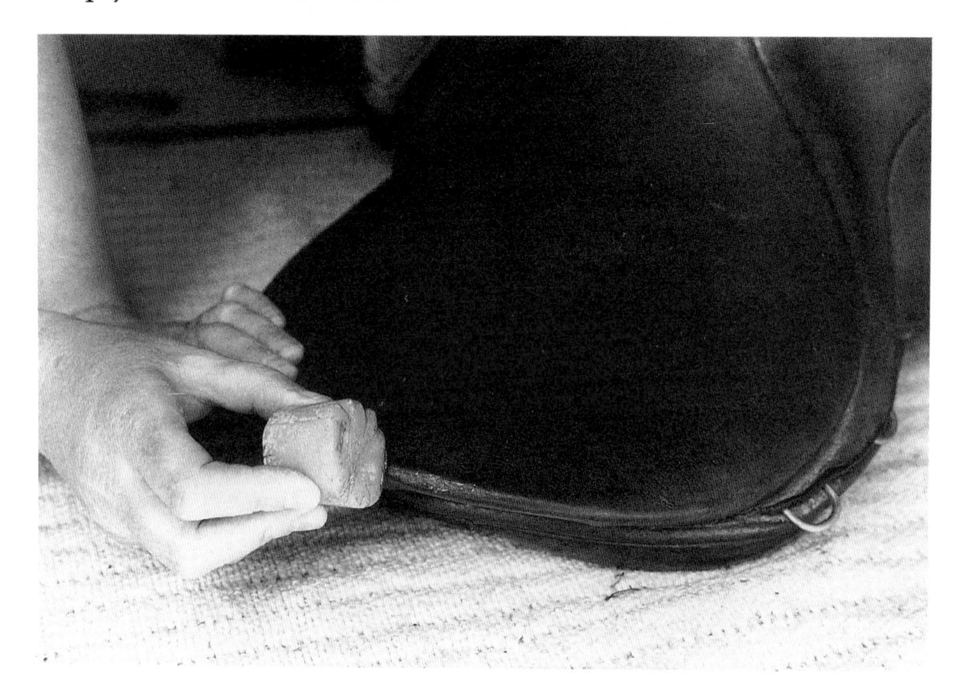

Fig. 4.16 By rubbing beeswax over the cantle repair, it will fill most of the indentations, and much of the damage can be disguised, especially after several coats of saddle soap.

This may sound an unusual type of repair, but I have used this method on my own hunting saddle and after a few coats of saddle soap it is, to all intents and purposes, invisible.

You may have the problem of part of the leather being missing. In this case you will have to cut a larger patch to go across the whole width of the seat with enough overlap to fold over the cantle (see Fig. 4.17). Glue the patch in place, having previously scraped the saddle seat to give a key to the glue. If possible leave it until the following day to allow the glue to dry completely, then thoroughly soak the part of the patch overlapping the cantle with warm water and pull it around the cantle. It can then be

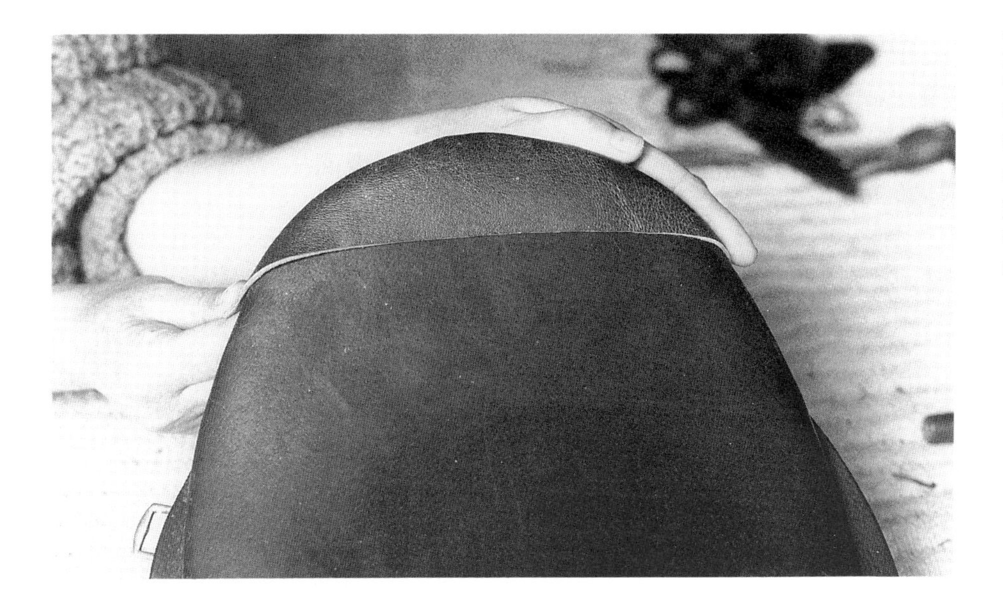

Fig. 4.17 A patch has been cut to cover the complete cantle part of the saddle seat. Once again the edges of the patch would normally be stained before gluing it onto the saddle seat.

secured in one of two ways:

1 Mark the leather where the lacing joins the panel to the seat top. Cut on this mark, allow to dry, then glue into position.
2 A far better result is achieved by cutting the long stitches which hold the panel to the seat top, thus exposing the back of the tree (see Fig. 4.18). With the patch still wet, pull it

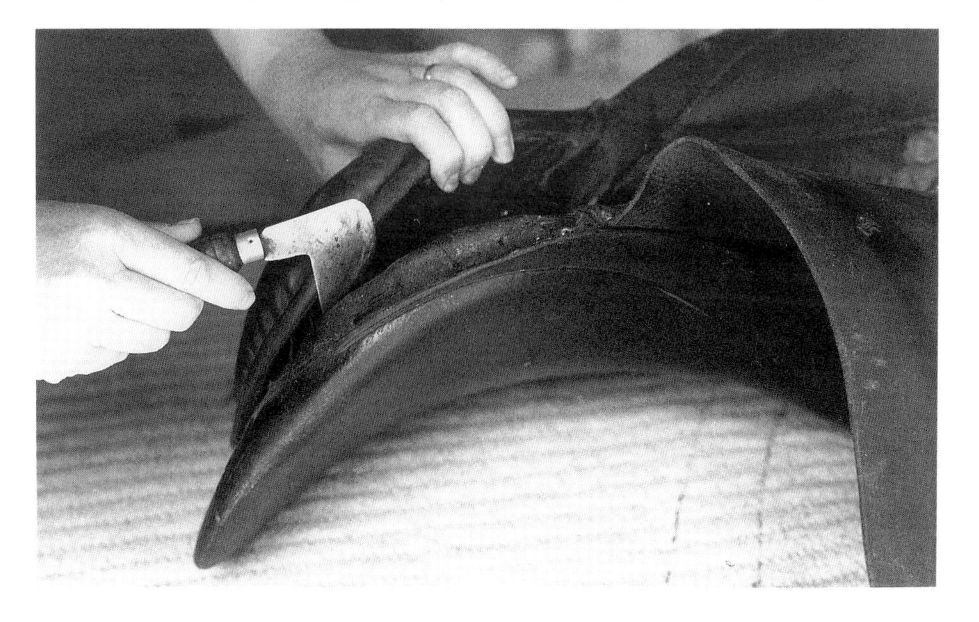

Fig. 4.18 The back of the panel is held onto the saddle with long stitches one to two inches apart. When the first couple of stitches are cut the rest will pull apart easily.

down as far as possible and tack it into position below the line of stitching you have just cut (see Fig. 4.19). This will secure the patch in place permanently but you will have to restitch the back of the panel with a curved needle.

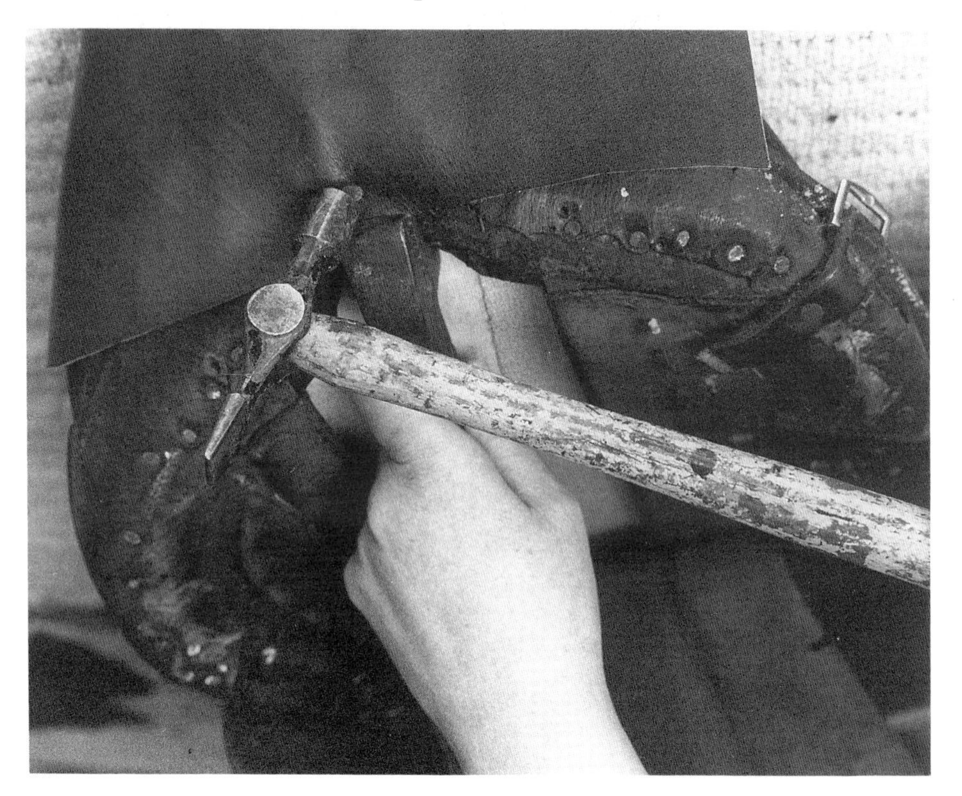

Fig. 4.19 With the patch glued into position, it can be dampened to make it pliable and tacked onto the saddle tree. Bull-nosed pliers with serrated teeth can help hold the leather in place, but, failing these, ordinary coarse pliers or pincers would be serviceable. Nail the patch as close as possible to the steel plate that reinforces the tree.

Splits in the seat

Small splits along the seam of a seat are very difficult to repair. Apart from completely reseating the saddle, the only alternative is to stitch a patch over the split, but this can be uncomfortable for the rider. It is probably best to leave a small split because it will not render a saddle unsafe (see Fig. 4.20).

Curved Needles

Up to now you have been able to effect all the repairs with straight needles because you have had access to both sides of your work.

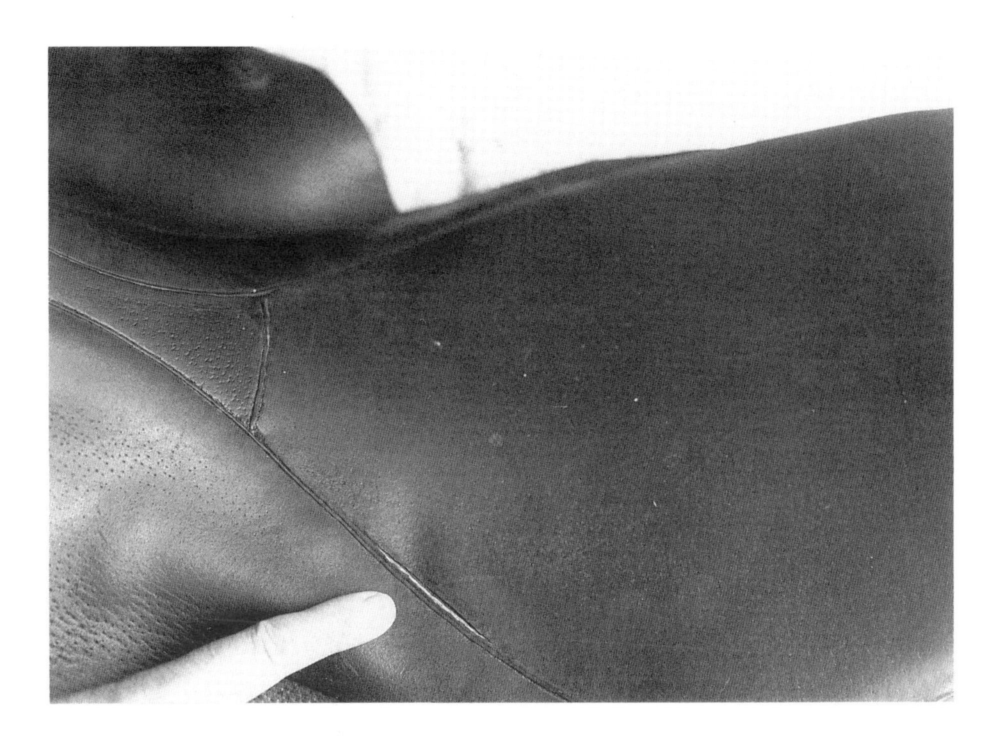

Fig. 4.20 A small split like this does not make a saddle unsafe and can be forgotten. This is my hunting saddle and I prefer to leave the split rather than patch it.

Now you will need to use a curved needle and an awl with a curved blade (see Fig. 4.22) which make it easier when you are only able to stitch from one side of your work.

Examples of jobs requiring curved tools are: pulling up or lacing the back of the saddle panel to the seat top; relining driving collars; fitting new leather tops to riding boots.

With a curved needle you will be able to stitch these examples by lacing the thread, (i.e. putting it in and drawing it out on the same side of the work.

Making a Curved Needle

You can either buy a ready-curved needle, or, to make you own, take the largest straight needle you have and, holding it with a pair of pliers, heat it in a gas flame until it is red hot (see Fig. 4.21). The shape can easily be formed by tapping the needle with a small hammer.

Using a curved needle to lace the panel into place

Thread the needle with double thickness, heavily waxed 5-cord thread. It is possible to buy special plaited nylon thread, for this

Fig. 4.21 You can either buy curved needles or make your own. I make mine by holding a large needle over a flame until it is red hot and then tapping it to shape with a hammer.

job, but the small amount you are likely to use would make it impractical. Traditionally saddlers used to make their own thread out of hemp, twisting several strands together by rubbing them along their thighs before waxing. The modern ready-made threads are, in my opinion, equal to, if not a little better than, these old-style handmade threads.

Thread the needle, and holding the two ends of the thread together, (i.e. double thickness thread) tie a large knot that will not pull through the lace holes. It will help if the stitching at the back of the panel does not need completely restitching, because the remaining sound stitches will hold the panel correctly in place. You will see that the stitches are quite long, perhaps between one and two inches, and are laced with one stitch in the panel and the next in the saddle seat, and so on. Start a couple of stitches back from those that have been cut or worn and follow the direction of the old stitches using the same holes. Unlike all other stitching so far, do not pull the stitches tight as you go, instead, leave a loop at each stitch you make (see Fig. 4.22). These will be pulled tight later because, by lacing with the stitches left loose, you will have more room to operate. Work your way around the saddle until you again overstitch two or three good stitches.

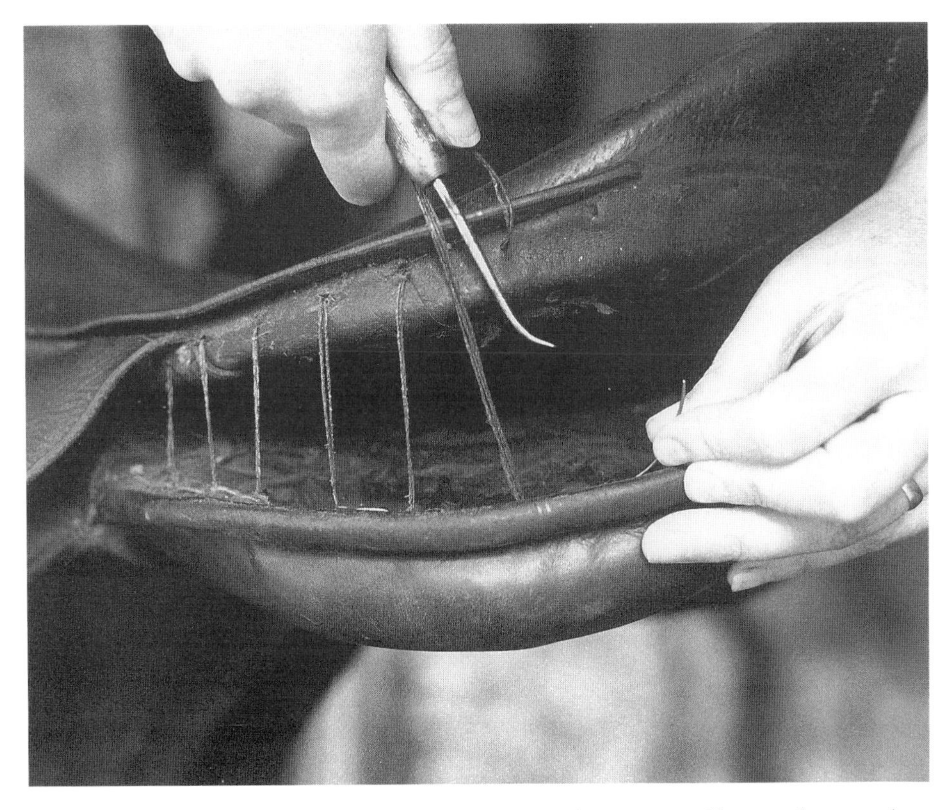

Fig. 4.22 To pull up the back of the panel, use double thickness, heavy thread and a curved needle and awl. Do not attempt to stitch tightly as you go round, but simply follow the line and holes of the old stitching and leave a gap as shown.

If you are lacing onto a cantle patch you will not have the stitch holes on the seat to follow, but you can easily use the marks on the panel as a guide.

To pull the stitches tight, use a strong tool to wrap the thread around and take up the slack at each stitch (see Fig. 4.23). Take care to pull the stitches along the line of the stitching and do not try to pull one part of the saddle to the other or you may well split the leather. When you have all the stitches tight, tie a large knot in the thread and tuck it away out of sight.

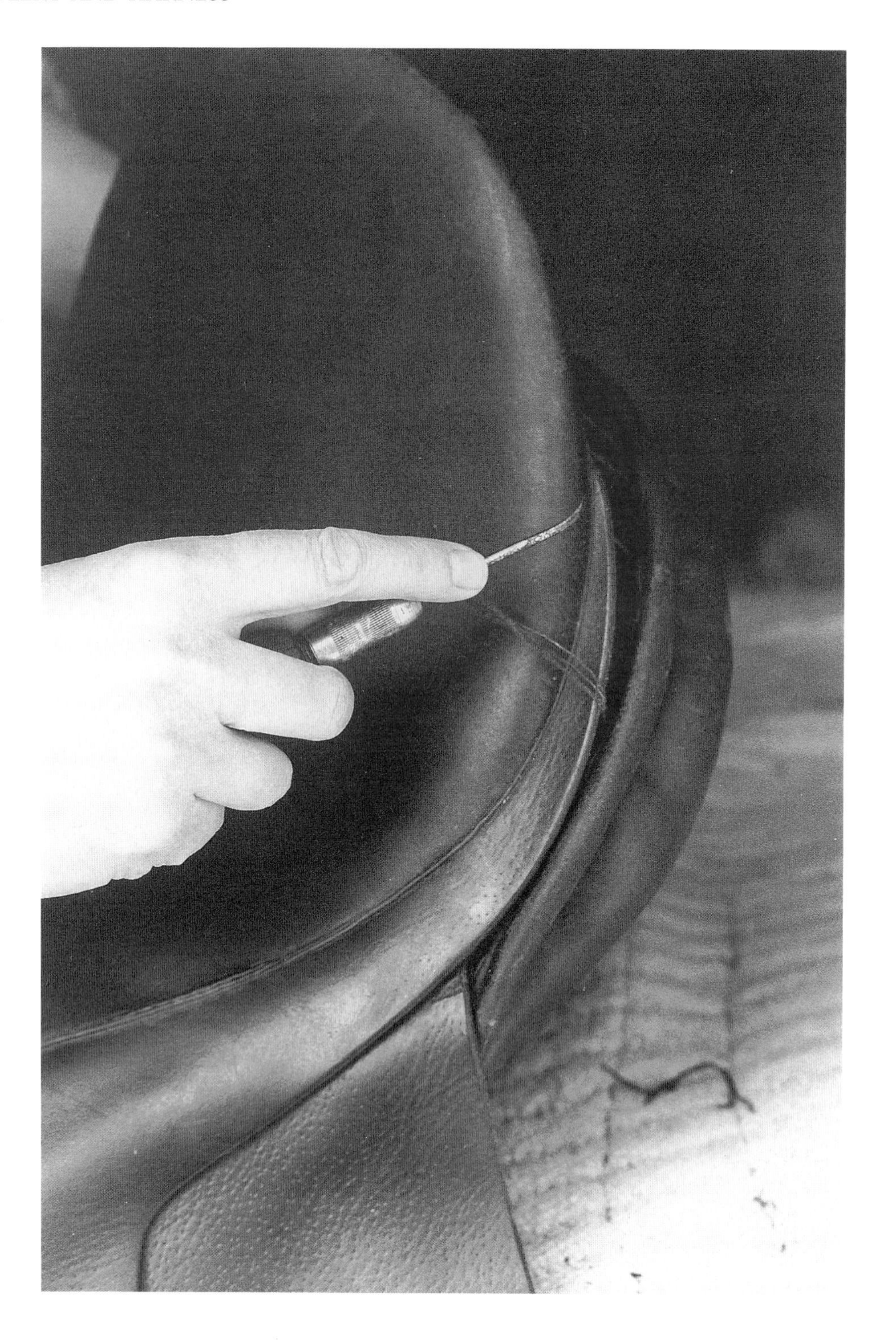

Fig. 4.23 With all the stitches in place it is time to pull each of them up tight in turn. Do this by wrapping each stitch around the awl blade several times and pulling, keeping the line of tension in line with the stitches so as not to split the leather, i.e. do not pull against the leather.

Fitting a Crupper D or Ring

To stop a saddle sliding forward on flat shapeless ponies, a crupper often proves very effective. Fitting a crupper D to a saddle can be achieved three ways.

1 Screwing a staple directly onto the back of the cantle (see Fig. 4.24). This method is easy and only requires a screw driver but the stress put on it can often cause the screws to pull out of the saddle tree.

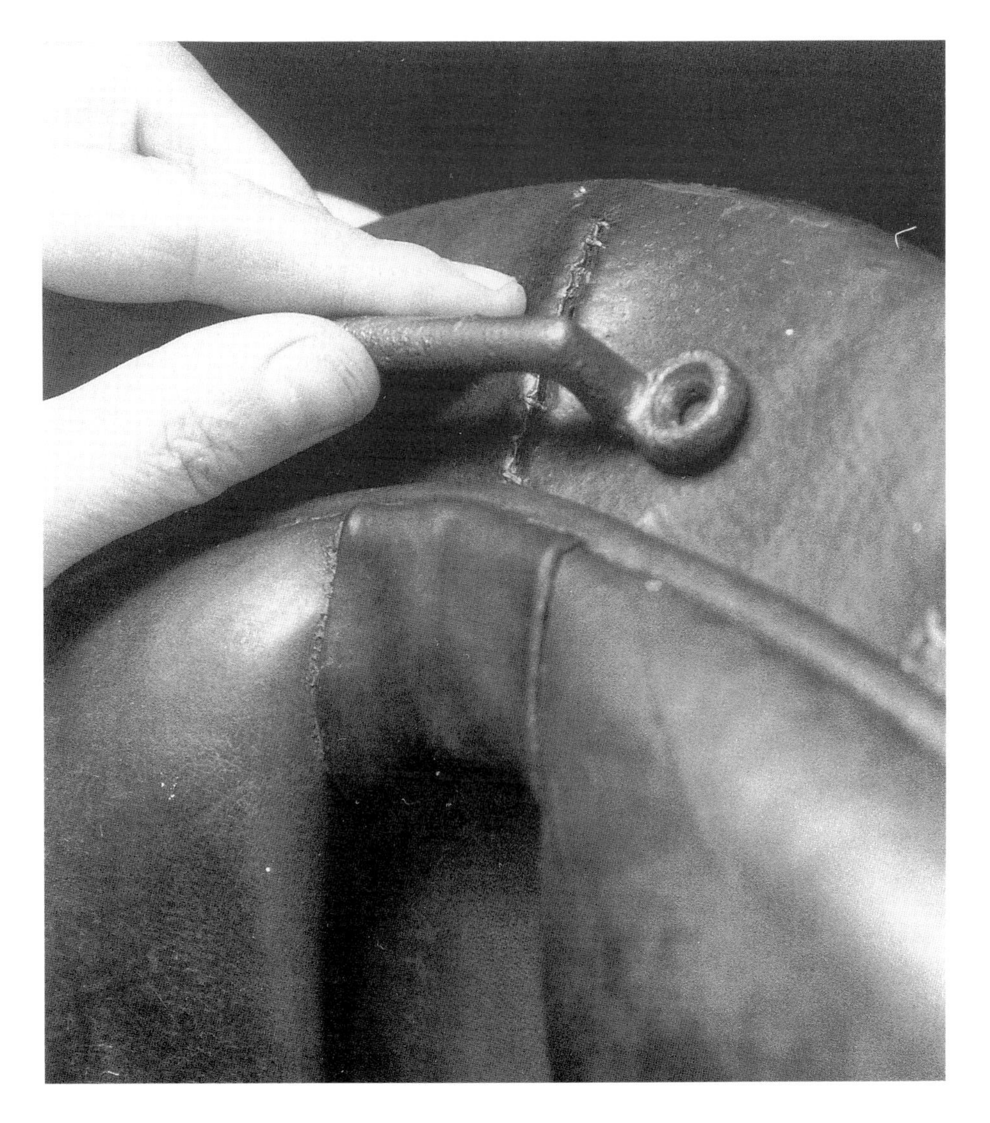

Fig. 4.24 A simple way of fitting a mounting to take a crupper is by screwing a staple onto the back of the cantle. This type of fitting is much in evidence on military saddles and on long-distance riding saddles used on the continent. A good way to secure panniers to the rear of the saddle is by fitting two staples like this to the back of the cantle at an appropriate angle to take your pack.

2 By dropping or taking down the back of the panel, a ring or D held in place by a piece of leather folded around it can be screwed onto the underneath of the saddle tree. The stress is distributed over a larger area and longer screws can also be used. You will then, of course, have to restitch the back of the panel as previously described.

3 Many small ponies are ridden with felt pads, and a crupper D or ring can easily be stitched onto the back of the seat, the stitches going through both the felt and the leather (see Fig. 4.25).

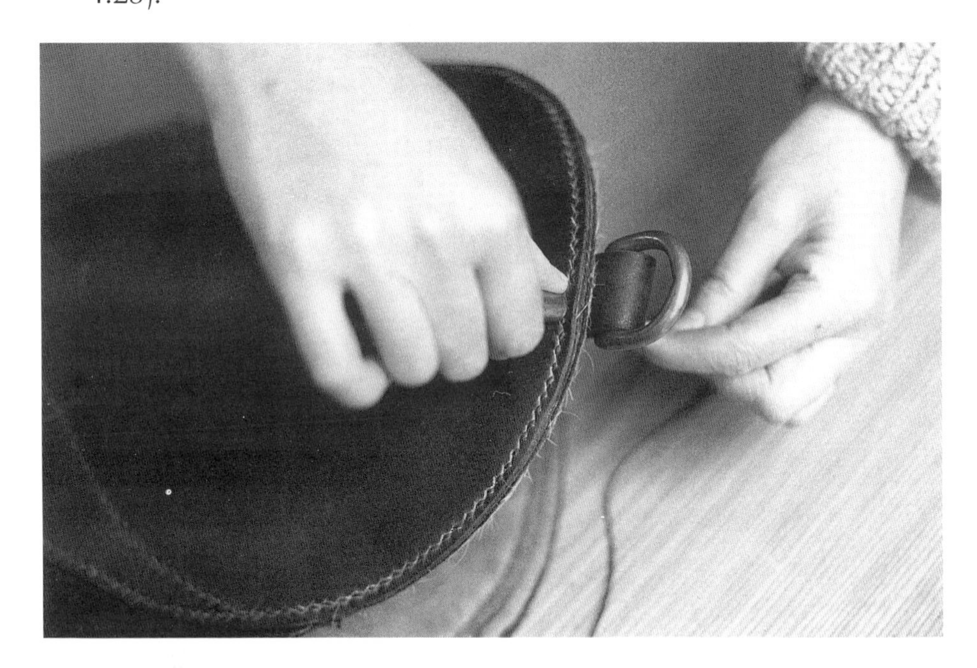

Fig. 4.25 Many small saddles and felt pads can easily slip forward when used on fat ponies. Stitching a D to a felt pad is an easy job and allows you to use a crupper to help keep the pad in place. Insert the D between the leather and felt and stitch through securely. As well as a D you can use a ring, or a buckle with the tongue taken out as in this picture.

Repairing the Saddle Panel

The most common repair needed on panels is the restitching of the panel ends. Being a straightforward stitching job it is covered in Chapter 2.

Repairing Tree Pockets

Occasionally the tree pockets may become unstitched and, although looking very unsightly, this will not actually cause harm or damage to the saddle when in use. These pockets were the first parts to be stitched on when the panel was new, and to restitch them as they were stitched originally is impossible, unless you completely remake the panel. If it is just the bottom of the pockets that need restitching, you can try to stitch the pocket to the panel via old existing stitch holes by slanting your awl, with the longest blade you have, from the pocket to the unflocked part of the panel. This sometimes works, but not always (see Fig. 4.26).

Fig. 4.26 The stitching at the bottom of the tree pockets can sometimes be accomplished by slanting the awl to reach the stitching on the inside of the panel.

Another method of repair is to remove the points pockets completely and stitch them onto a large patch of strong but thin skin (see Fig. 4.27). Then, after sliding the tree points into place, lace the patch onto the panel using a curved needle.

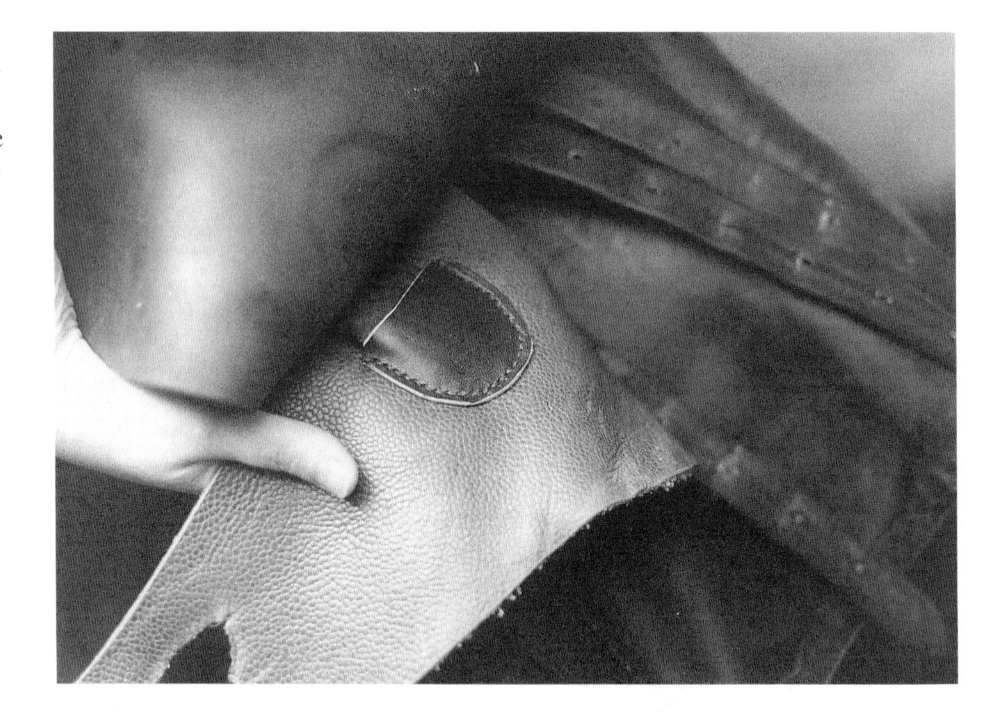

Fig. 4.27 If the pocket either needs renewing or restitching completely, then, first, it needs to be stitched to a larger piece of leather which is positioned, trimmed to size if necessary, and then laced into place.

Panels: Leather Lined

Panels wear most frequently across the top of the knee rolls. If it is only the lining of the knee roll itself that is worn, then it is possible to glue and stitch a patch across the panel (see Figs. 4.28, 4.29 and 4.30). A great disadvantage to this is that it is really impractical to use the saddle again without a numnah or cloth underneath, because the horse's hair and sweat will work their way under the patch.

An alternative to using a patch, especially when other parts of the lining are showing signs of wear, is to reline the panel completely. However, to replace a leather lining in the style of the original is a highly specialized job and one that, like a new saddle seat, is very rarely justified on economic grounds.

Overlining

An overline *is* perfectly feasible for this repair. This method can be carried out equally well on a leather-lined or an older material-lined saddle. Unless the shape of the panel is completely

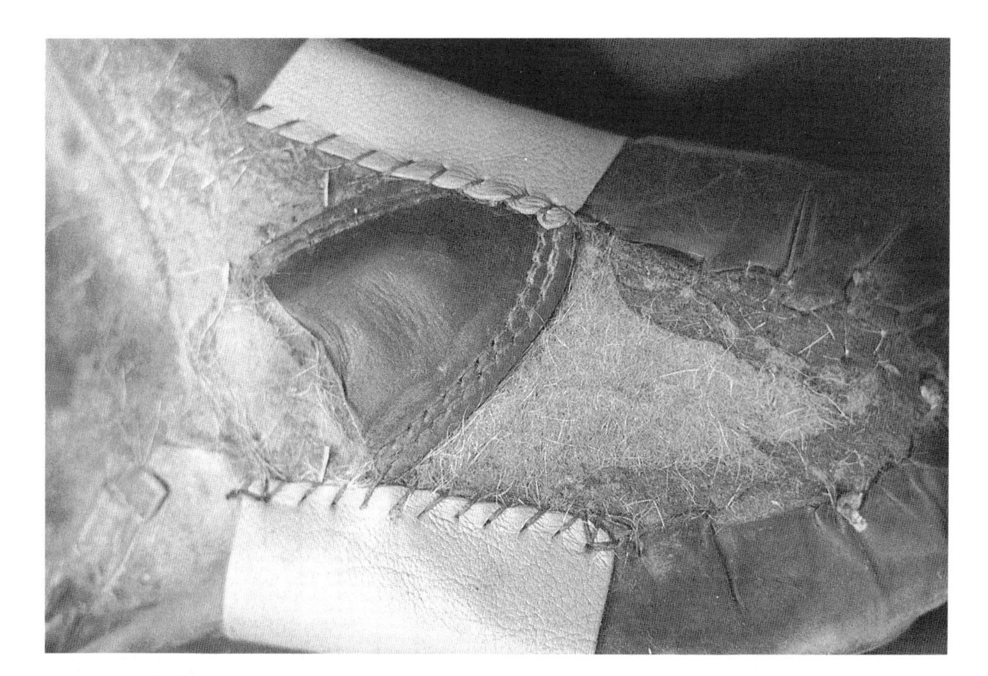

Fig. 4.28 Some saddles may have a solid felt panel covered with a thin hide. This is such a panel with a repair prior to staining.

Fig. 4.29 The patch on this panel was stitched into place with a curved needle.

flat, like many old surge-lined saddles, it is easier to use either two pieces of leather joined along the middle, or three pieces, using the third piece for the channel. With many old flat-paneled saddles, it is possible to overline with a single piece of hide.

Once you have decided whether to use one, two or three pieces, select a thin but strong hide (see Chapter 1 for types of hide).

Fig. 4.30 The awl used for this repair has a curved blade to facilitate stitching onto a flat surface.

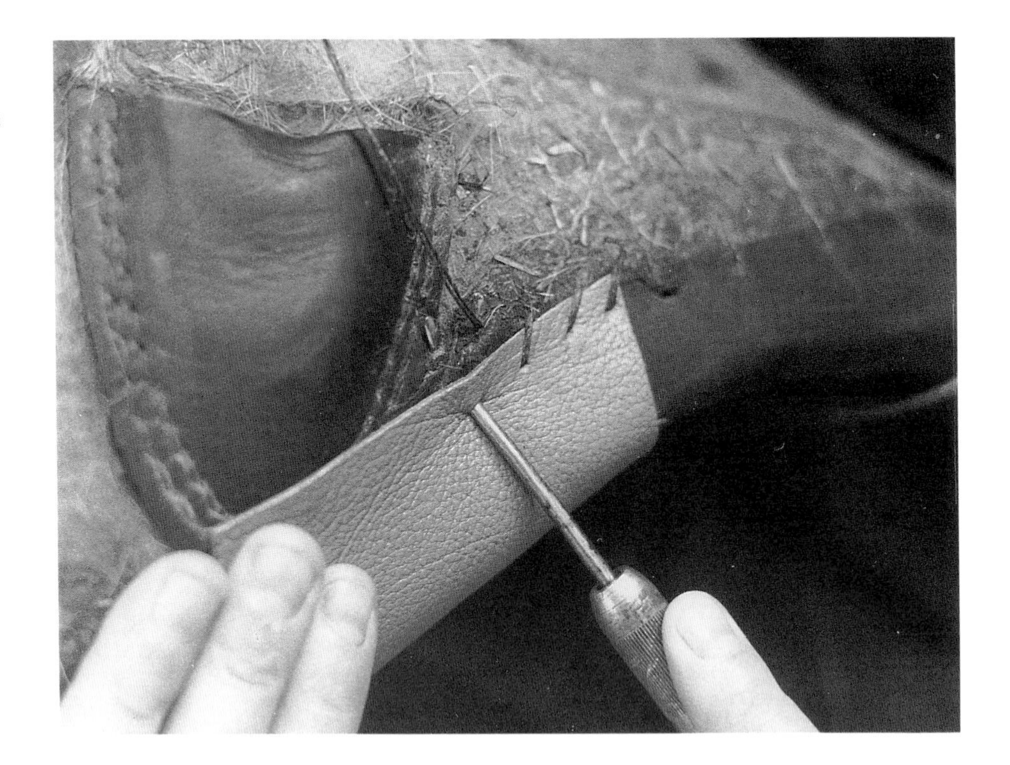

Keep it secure while you stitch it into place by gluing the new lining onto the saddle, using a solvent-based glue for sticking leather to leather, or a rubber-type glue for sticking leather to material.

You will need to stitch round the edges and, in the case of a quilted panel, to requilt following the old square pattern. Allow an overlap of ½ in and, as you stitch, following the old stitches round, turn the overlap underneath the lining, so the raw, cut edge of the new lining will be hidden between the old lining and the panel. This is far better than stitching first and cutting off any excess afterwards, thus leaving a raw edge. A straight needle is correct for stitching most parts, but, around the thicker padding under the saddle seat, you may well find it easier to use a curved needle because the skirts and cantle will prevent you from stitching in a straight line.

Remember that, when pulling leather to shape, if it is dampened with a little water it becomes pliable, and therefore easier to stretch to the desired shape.

Panels: Material Lined

Many old saddles, still in regular use today, were made with either a woollen or linen lining. These saddles should always be used with a cloth underneath so that any sweat can be absorbed by this and not directly into the saddle lining.

If these linings become worn they can easily be re-covered, either with leather as previously described, or with an old, heavy woollen blanket or rugging. You should be able to stretch this woollen material across the entire panel and replace it in one piece. Again, a little Copydex-type glue around the edges helps to keep it in position while stitching. With rugging it is imperative that the edge is tucked under as you stitch around the panel, so no bare edge is left to fray (see Fig. 4.31). Surge-lined saddles will certainly need to be requilted through the panel to keep the material in place (see Fig. 4.32).

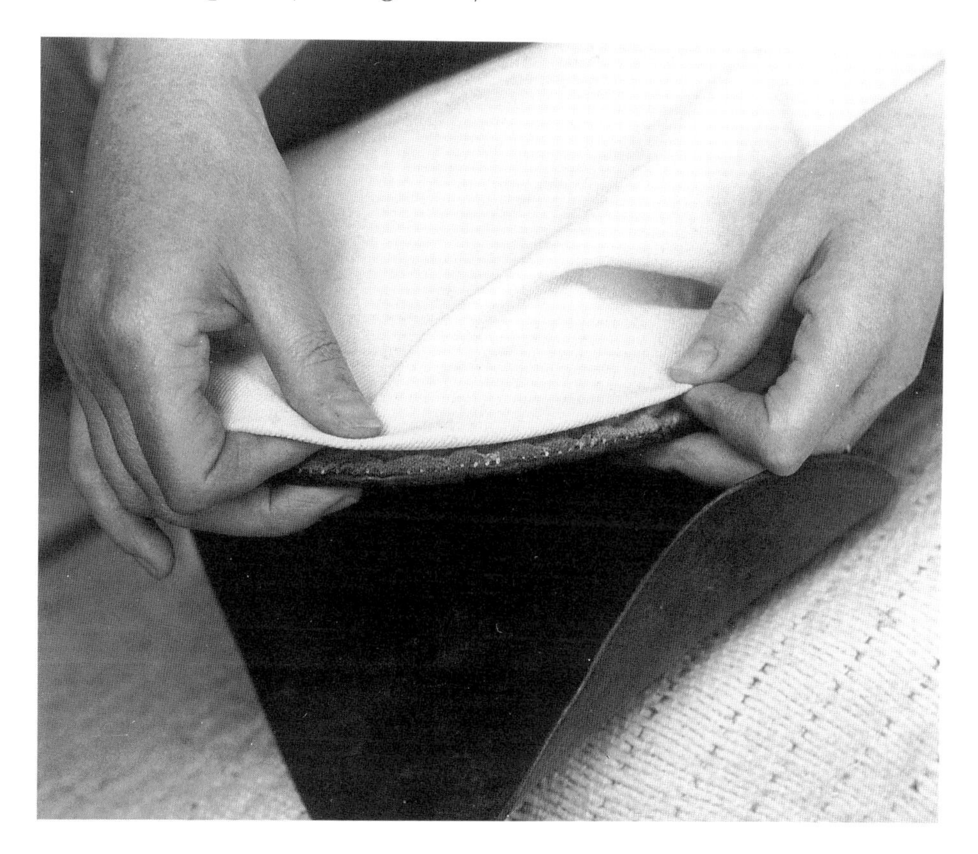

Fig. 4.31 When relining with material always fold the raw edges underneath then pop stitch around the edge before quilting.

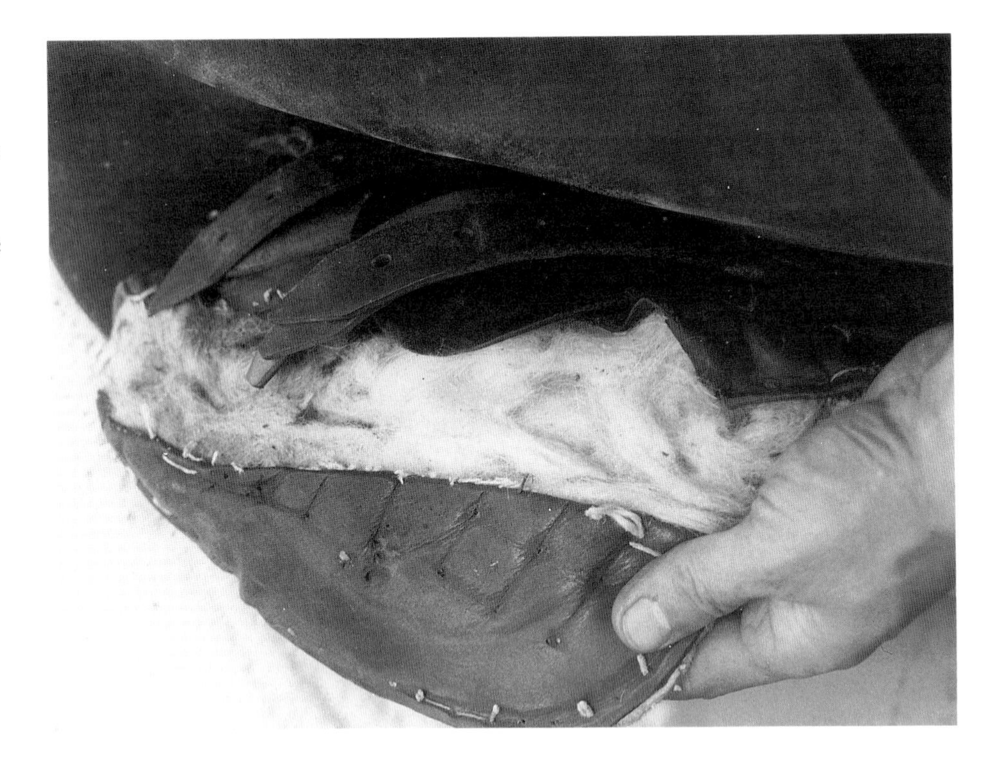

Fig. 4.32 On this old style, quilted-panel saddle, the panel has become unstitched, but is not in such a drastic condition as it may first appear. By restitching the two ends of the panel together and requilting, the panel will be serviceable once again.

With some old saddles and most side saddles the material lining of the panel is made up of two pieces each side of the channel. The weight-carrying bearers – the heavily padded parts of the panel running along the length of the saddle each side of the channel – are covered with one piece, with the quilted side pieces covered with a separate piece of lining. If this is the case, feed the material for the quilted part underneath that of the bearers and stitch into place (see Fig. 4.33). Unless the lining is totally destroyed (see Fig. 4.34) you should be able to follow and re-cover the old lining in its original style.

Requilting

To requilt use a single needle and heavy white thread, and follow the old stitches approximately one inch apart. Use a larger stitch on the leather side and a much smaller stitch on the lining side, just catching-up the lining. The finished job will have a 'squared' effect on the leather side, and stiffen the panel quite considerably.

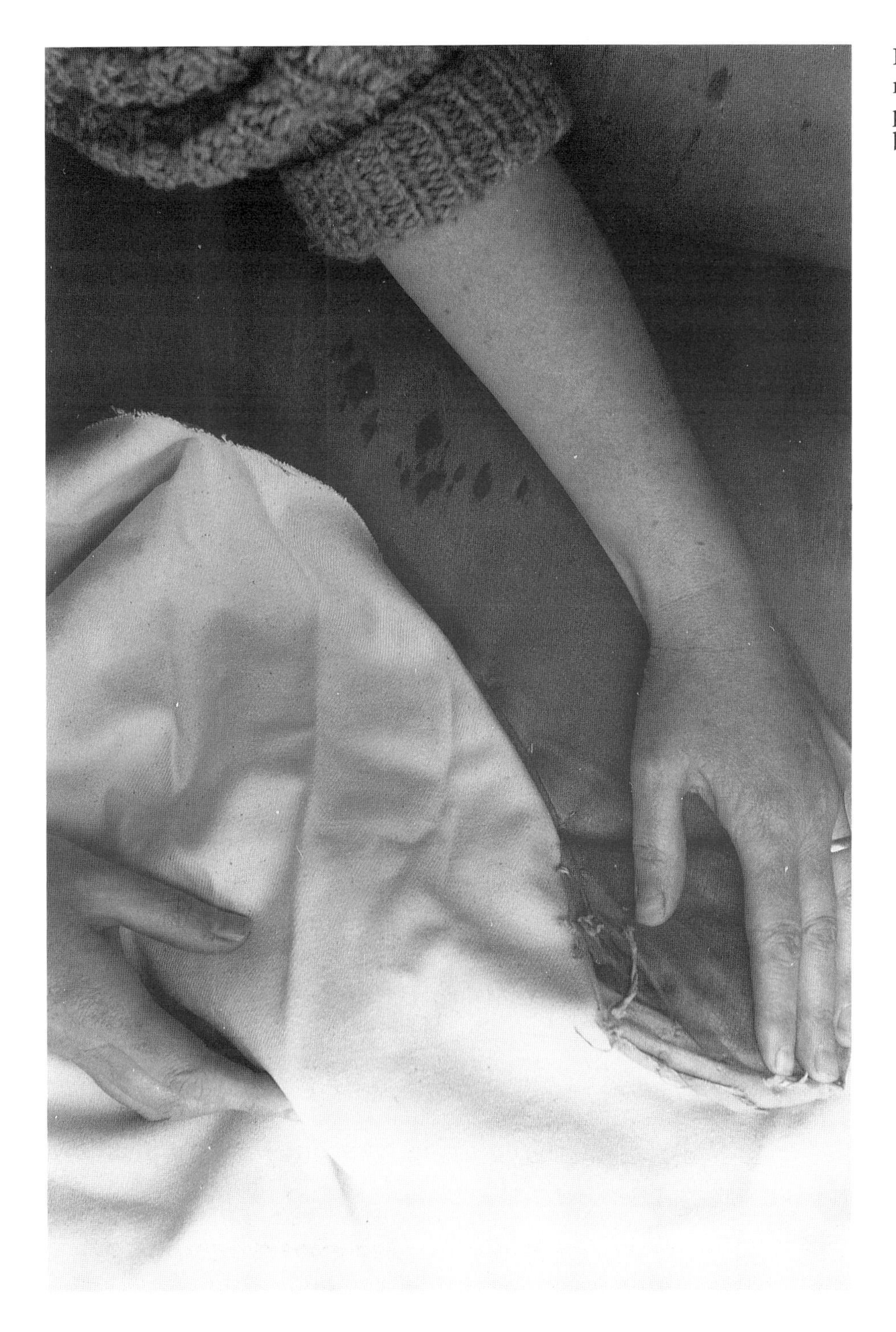

Fig. 4.33 Feeding the material for the quilted part of the lining under a bearer.

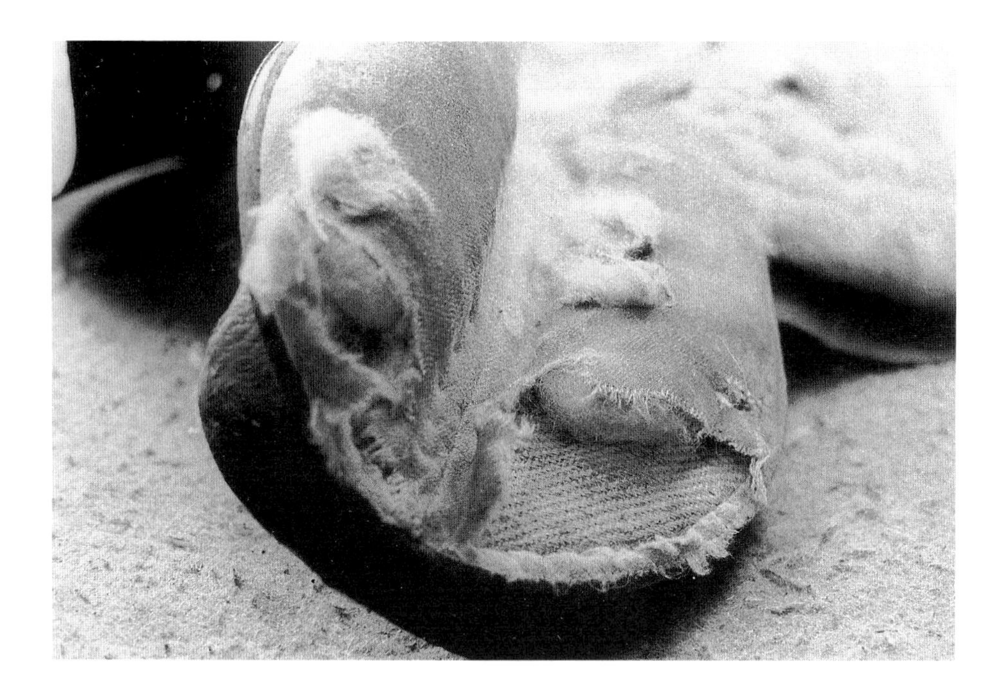

Fig. 4.34 A totally destroyed serge lining.

Knee Rolls

The modern trend towards larger and longer knee rolls often means that, instead of being made up as part of the panel, they are made separately and then added. This often results in the stitching becoming weak and requiring repair. Many of the large type of knee rolls are laced onto the panel using a curved needle, and it is an easy job to restitch them. Simply thread a curved needle with double thickness thread and follow the old stitch holes. Although the holes are already there it is helpful to reopen them by first using the curved awl. You may be able to stitch them tightly as you go along, but if you find difficulty in holding the thread tight against the stitches already made, then stitch loosely and pull up at the end.

Restuffing the Panel

This is one of the most skilful jobs a saddler has to do. Both sides of the panel must be equal, it must be level, it must not be lumpy, and, finally, it must fit the horse's back. Restuffing is not therefore, a job to be recommended for the beginner. I have seen many restuffs that have left a lot to be desired. The job should be done by a saddler who knows his horses and understands the need for a rider to be properly balanced in the saddle.

However, there is a way to lift a saddle slightly without having to restuff with flock, and that is to use fillets of felt. It is the only way I have found to adjust the balance on felt-paneled show saddles used for children's riding classes. Two lengths of felt are cut just shorter than the length of the panel and about two to four inches wide. This is then either slipped underneath the panel next to the saddle tree via the channel, or the front of the saddle is taken down and the felt tacked onto the saddle tree. Although this is very simple, especially if the former method is used, it can prove to be very effective. Most native ponies have no wither and the extra thickness of felt can make all the difference.

Broken Trees

Many saddlers will claim that it is either impossible or unsafe to repair a broken tree; I disagree. I have successfully repaired many saddle trees and after lending my own saddle to a friend, who's horse promptly reared and came down on it, have also regularly ridden on just such a saddle. It is not, however, a job for the inexperienced. I remember, as an apprentice, being told to take a saddle needing repair to the local blacksmith and watching helplessly as it caught fire at the end of a welding torch.

There are basically three ways tö repair a saddle tree.

1 Stripping the saddle back to the tree and gluing and taping the broken woodwork.
2 Reinforcing the broken woodwork with specially made metal plates welded into position.
3 Welding the arch. When only the arch has broken, a good welder can, with care, effect a strong and safe repair.

The first two repairs are, frankly, out of the question for an unqualified saddler. If your saddle tree is broken along the length of the tree (see Fig. 4.35), then visit a saddler and ascertain whether or not he is happy to tackle such a repair and if he can offer a guarantee with the job. If only the arch has broken, you can, as stated, repair this with the aid of an experienced welder. A good welder will know the strongest rod he can use in relation to the thickness of metal to be repaired.

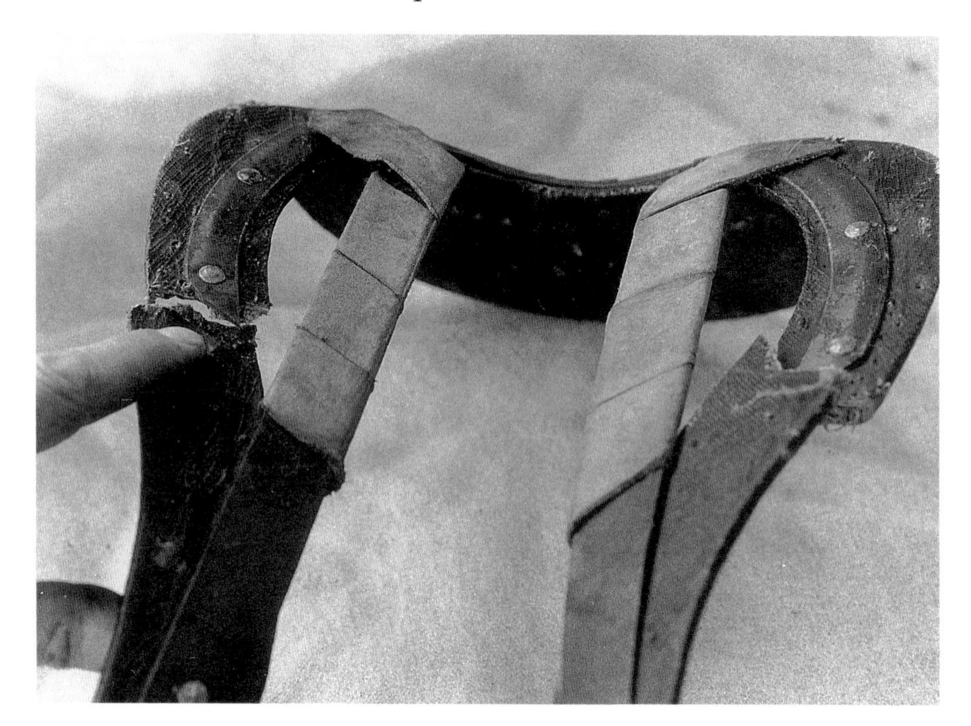

Fig. 4.35 The break in the wooden tree can be seen at the point where the metal reinforcing piece finishes. A break of this extent is not likely to happen during normal riding activity; in this case the horse fell over backwards on top of his saddle while jumping cross-country. The two flat pieces on the inside of the tree are the leather-covered springs.

First drop the front of the saddle and remove the tacks from the piece of thin hide that runs along the length of the channel covering the tree. You will now see the extent of the damage. The metal arch of a saddle will usually break in the middle (see Fig. 4.36).

The photo shows that there had been a previous attempt to repair the tree using a thin metal plate held in place with some rivets. I say 'attempt' because this type of repair will never be successful. The correct way to effect this repair – and the way I tackled this particular saddle after first removing the old plate – is to weld it. Start by thoroughly soaking the front of the saddle

Fig. 4.36 This saddle was brought into my workshop with some movement and a squeaky noise in the front. Once I had stripped it down the horrific evidence of a previously attempted repair was all too apparent. Instead of welding or fitting a new gully plate, a thin plate had been crudely nailed into place. After I had removed the offending plate, the break in the tree was welded using high tensile rods, and the saddle is still giving good service. Another point to note in these pictures is the dreadful practice of nailing the narrow front girth strap directly to the saddle tree. Instead of a single web passing completely over the main saddle webs, a scrap of web is held in place with half a dozen tacks. The amount of time and web saved is minimal compared to the extra safety a proper job would have given. But when the saddle is put together this shoddy factory practice cannot be seen.

in a bucket of water. Immerse the saddle until the leather and webs are well and truly sodden. This will help prevent the heat of the welding torch from burning the webs onto which the seat is built, but extreme care must be taken, so before the welding starts, make sure there is a full watering can of water within arm's reach. Immediately after each length of weld, soak the arch with water from the watering can, so reducing the heat. If the webs are burnt through the seat may never be the same again.

When the weld is completed leave the saddle to dry out in a gentle heat, then tack the leather covering the underneath of the tree back in place and pull up the front of the saddle.

With many jobs, I have mentioned that they are invariably easier than they look at first. Repairing a broken arch is the reverse; one false move and the damage could be irrevocable.

5

RUG REPAIRS

New Zealand Rugs

The repairs that saddlers probably like doing least are repairs to New Zealand rugs. It is very rare for the rug to be clean, it usually arrives straight off the horse, and, often, just picked up off the muddy field after falling off the horse.

Many of these rug repairs can be done without a heavy duty sewing machine and they are divided up as follows.

1 Repair and renewal of leg, surcingle and chest fittings.
2 Patching tears in the canvas. A small tear that is not repaired at once can very easily become a major rug-remaking job.

Let us deal with fittings first. Nowadays nylon web is used for the fittings, and has replaced the traditional grey chrome leather mainly because it is cheaper.

Leg straps

There are several designs of leg strap; some are stitched onto the rug, and some can be removed completely. All, however, should clip onto the back of the rug after going around the hind legs. Legs straps should clip on near side to near side and off side to off side, with one looped around the other (see Fig. 5.1). They should not be fastened, as some people fasten them, near side to off side and off side to near side.

If straps have become unstitched from the rug it is a fairly easy repair job because you can restitch them using the old holes. It

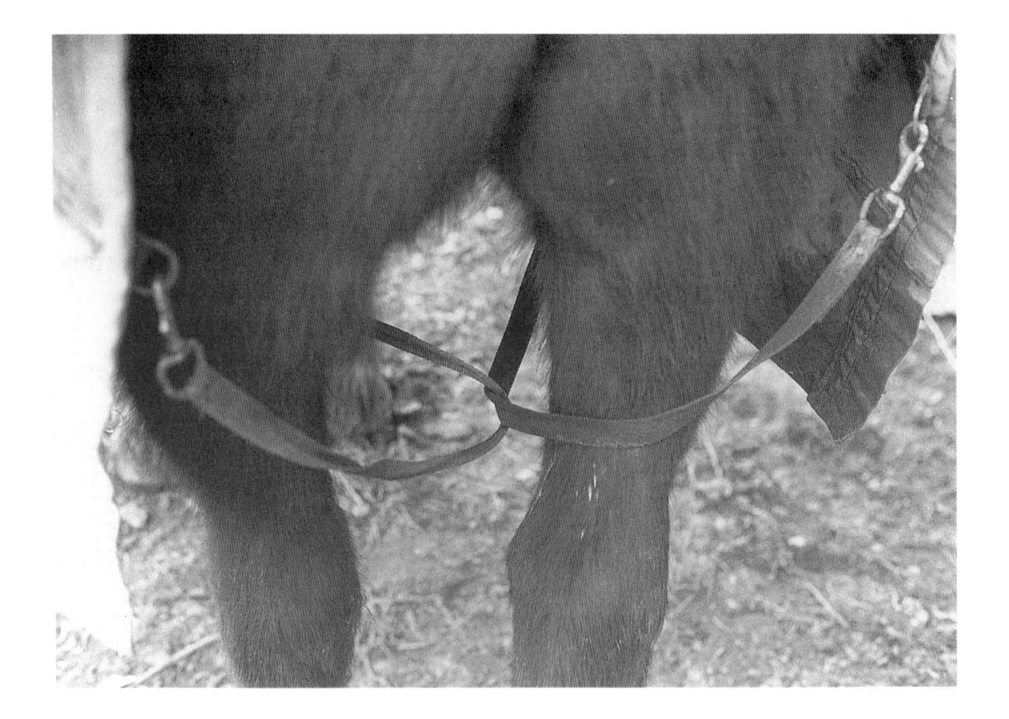

Fig. 5.1 The leg straps on a New Zealand rug should always be looped around each other rather than crossed in order to keep the straps slightly further from the horse's legs and so reduce the stress put upon them.

does not matter if they were machined on in the first place; just use two rows of holes.

The clip might have broken off or become so rusty that it cannot work. Always keep clips with moving parts, namely the trigger type, oiled. Any oil will do, even a leather oil. If a new clip has to be fitted, you have the choice of using the old stitching holes if the stitching has come undone or, if the strap has broken off or was riveted, to re-mark with a pricking iron, approximately six or seven holes to the inch and hand stitch on.

If the whole leg strap needs to be replaced, I suggest that a nylon web strap is used. It would be slightly easier to use a sewing machine for a webbing strap, but hand stitching will do the job well. You cannot, of course, mark the stitches on webbing with a pricking iron, but you should be able to use the weave of the web as a guide.

Surcingles and breast straps

With the improvement in New Zealand rug design – namely the extra deep pattern rugs and self-righting types with a drawstring at the rear – many of the better rugs no longer have surcingles. My present rug does not have one, and it seems to stay reasonably straight in use, but if a surcingle is used, it is far better to have slots in the side of the rug for it to slide through, rather than encircling the outside and gathering the bottom edge of the rug up under the horse's armpits. Again, if they have chrome leather fittings that require restitching then use the existing stitch holes. If replacing with web it will need to be hand or machine stitched.

To make up a double thickness web strap without a heavy duty sewing machine is a problem, although a suitably adapted domestic machine might do the job. It will not, however, be able to stitch the made-up strap onto the rug; it is a hand stitching job if a heavy duty machine is not available.

Many lightweight domestic sewing machines can be uprated with a coarser feed dog and foot, and thicker needles. A good sewing-machine mechanic should be able to advise you. Three addresses are given at the end of this book in the Appendix.

Buckles and Clips

ROKO, OR TONGUELESS, BUCKLES

Some people say that roko buckles slip, and they will, if they are put on upside down or used on single thickness web. I have two complete sets of driving harness with roko buckle fittings, and I do not have trouble with slippage. The buckle must, however, be the right way up. the wider bar, together with the moveable round wire loop must be at the top of the buckle when it is done up (see Fig. 5.2).

CONVENTIONAL TONGUE BUCKLES

These are normally for use with leather fittings because it is a straightforward job to punch out holes to accommodate the buckle tongues. If they are used in conjunction with web, metal eyelets should be inserted in the web.

Simply burning a hole through the web using a very hot nail, or something similar, will suffice for lightweight jobs, but not for straps that are going to be put under a lot of strain. After a very short time the sealed hole will expand and start to fray. For

Fig. 5.2 A roko buckle has a top and a bottom, a simple fact some manufacturers overlook. The wider bar together with the round wire loop *must* be at the top when it is buckled up.

something as heavy as a New Zealand rug, very tough web indeed would have to be used if the holes were simply to be burnt in.

METAL CLIPS

These fittings are used only on the surcingle in conjunction with a narrow metal slide adjuster. They could be replaced by more conventional fittings if needed.

NYLON CLIP-IN FITTINGS

Clip-in fittings are completely weatherproof and generally strong if well made. They are used more frequently on lighter rugs, such as day rugs and anti-sweat rugs, because they can be washed in a washing machine without any ill effects. Replacement clips are readily available at camping shops because they are widely used on rucksacks (see Fig. 5.3).

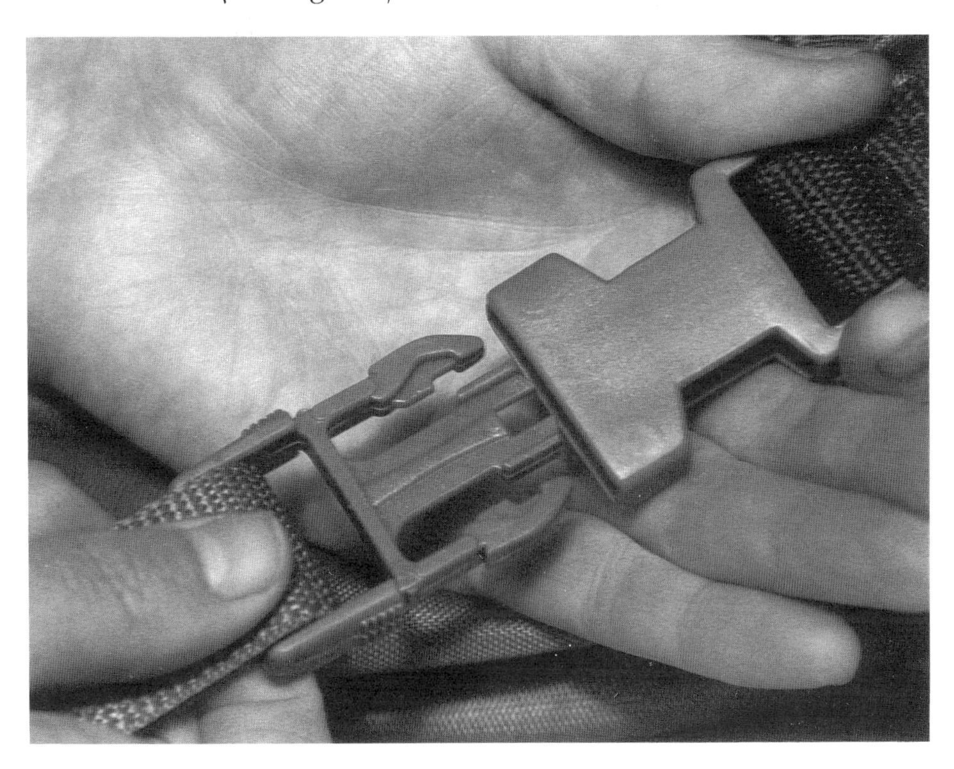

Fig. 5.3 Metal or plastic clips are now often used instead of the traditional leather buckle and point fittings on rugs. The great advantage with them is that they can be safely washed on the rug and do not need any oiling.

Oiling fittings

When replacing leather fittings on the surcingle and breast part you can substitute ordinary vegetable tanned leather for the grey chrome. It will not, perhaps, last quite as long, but if kept well oiled could still outlast the rug.

Grey chrome tanned leather is often marketed thus '. . . with weather-proof chrome straps. Never needs oiling'. Rubbish. It cracks and snaps like any other leather, so oil it with good leather oil regularly, and remember to put a few drops on the leg strap clips.

Repairing canvas and nylon

Without a heavy-duty sewing machine, any large amount of stitching on these materials is going to be a problem. As mentioned before, a domestic machine can be used if it has the stronger attachments. It will normally cope with two thicknesses of material, but with the addition of a heavy lining and reinforced parts around the neck and wither areas of the rug, it just will not manage.

Hand stitching is slow but can produce a very good repair.

Remember that large repairs usually start as small repairs that

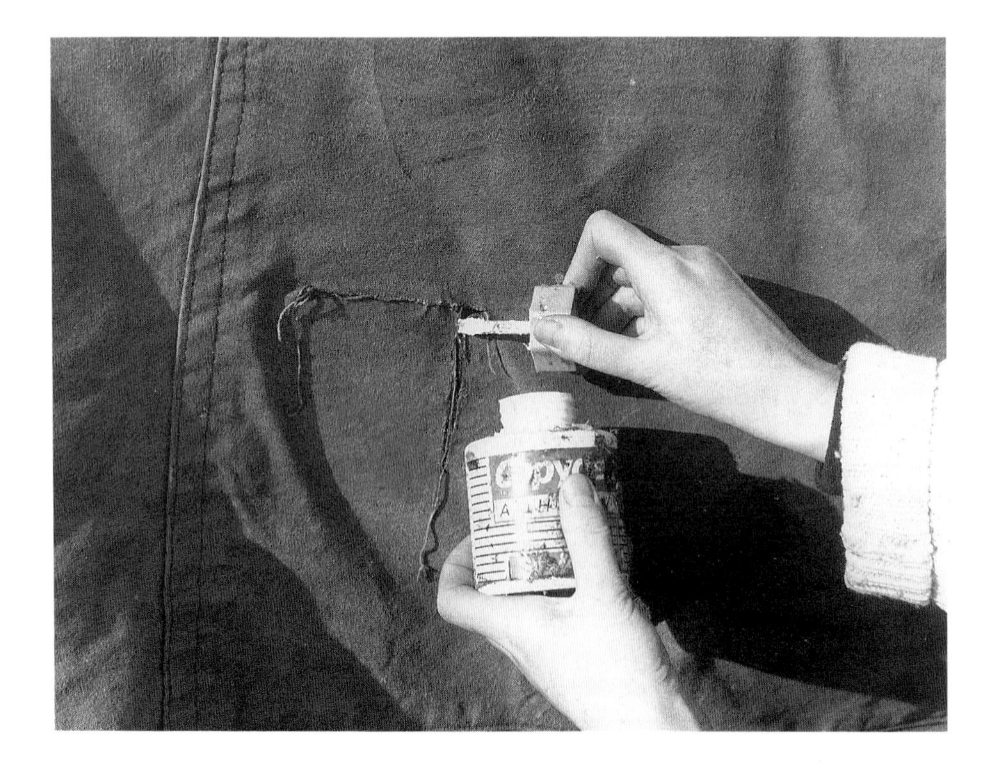

Fig. 5.4 When patching a New Zealand rug, glue a patch underneath the tear with a latex adhesive. This type of glue is waterproof and penetrates the weave of the fabric.

Fig. 5.5 If the rug is dry the repair can be carried out with the rug still on the horse. A quick patch in the field can save a more extensive job later.

were never mended, but a simple repair to a small tear can be effected either with the rug on a bench, or even while it is still on the horse.

At the start of the rugging up season have ready several pieces of strong cotton, linen, or even scraps from an old New Zealand rug. Keep these handy with a small pot of latex glue; ideally a pot with a brush in the top. At the first sign of a small tear, apply glue to the whole of one side of a patch and the *underside* of the rug around the tear. Allow the glue to go off for a few minutes then apply the patch inside the rug to cover the damage, glue-side up. Press the glued torn section of the rug down onto the patch, and keep pressing till the glue sets and everything binds together. If no part of the rug has actually been torn off, you will not see the patch at all. This can be further strengthened by using a hammer on the patch and so drive the glue into the weave of the patch. It is advisable, however, to do this with the rug removed from the horse! So instead of a small hole maybe two to three inches long that can catch on any protrusion and rip the entire side out of a rug, you have effected a tidy, sometimes almost invisible, repair (see Figs. 5.4 and 5.5).

When patching larger areas that are going to need stitching, always use latex glue to hold them in position and, again, use a hammer to aid glue penetration.

Anti-sweat Rug

The cotton-mesh anti-sweat rugs or sheets, look and work like string vests. In my experience, however, they are almost impossible to repair once they start to break up.

There are two tips which will help prolong the life of anti-sweat rugs.

1 Wash them regularly to reduce the time the sweat is in the material, or the sweat will destroy it.
2 Always buy an anti-sweat rug about a foot shorter than any other rug you would buy for a particular horse or pony. This is because when in use, and especially when travelling, the rug stretches. Once it is over the curve of the rump and a horse

rests his quarters against the stable or horsebox, the rug is trapped and, consequently, it is stretched to the point where the meshing breaks and the holes get larger.

If you need to replace the fittings on the front of an anti-sweat rug, try to use the nylon mountaineering type because these wash well.

Day/Travelling Rugs

Because these rugs are generally kept for travelling and for best, they tend to last a long time. With careful use you may never have to repair the material and, to aid rug preservation, check for any protruding nut and bolt heads in the horsebox or trailer, on which an animal could lean while travelling (see Fig. 5.6).

Storage can also be a problem; I have lost several rugs through moth damage. Highly scented, but cheap, bars of soap kept in the rug chest evidently have too powerful a smell for the moths and they leave well alone.

Fig. 5.6 Always check a horse box for any protruding nuts, bolts, screw heads, or anything that can snag and tear a horse rug. It only needs a single bolt head in the wrong place to cause damage.

Every so often these rugs need washing and this presents no problems if they have nylon breast fittings. Many of the better rugs, however, have leather fittings. The leather could be oiled thoroughly before and after a cool wash, to protect it, but a better bet would be to carefully cut the leather fittings off before washing, pick out the old stitches, and restitch them once the rug is clean. I know this sounds long-winded, but, once leather has gone through a washing machine, it is never the same again.

6

DRIVING HARNESS

Much of the information in previous chapters will also relate to driving harness. In general the principles of cutting, preparing, marking and stitching remain the same. I will, however, cover a few jobs that require additional information regarding the techniques used.

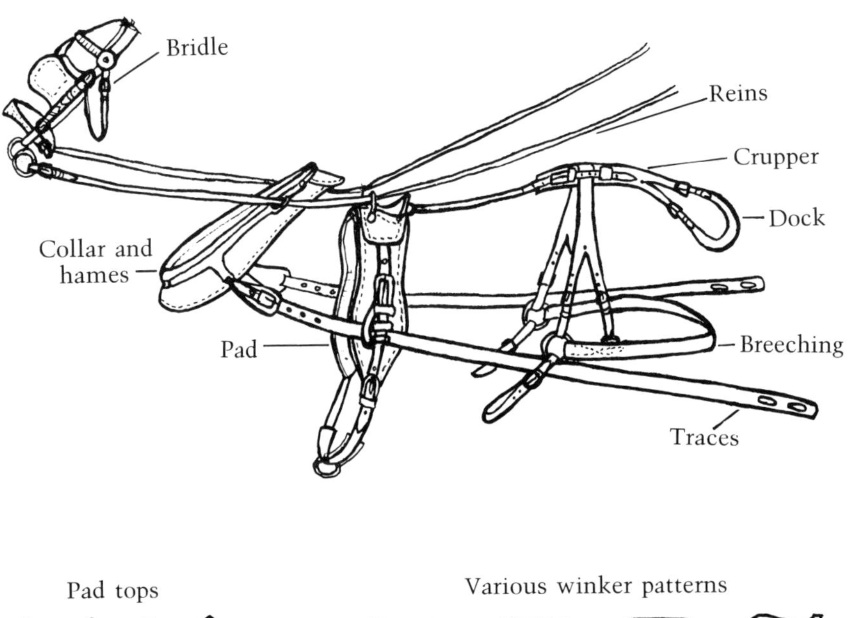

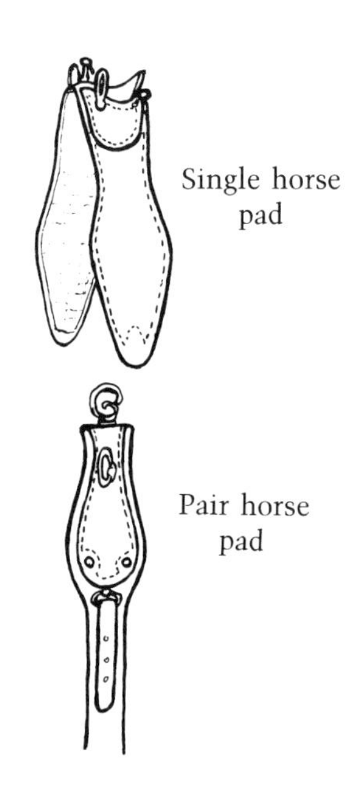

Fig. 6.1 Parts of driving harness.

Bridle

The two parts of a driving bridle most likely to require repair are the billets on the ends of the cheek pieces, and the winker stays.

Cheek billets

If the billets need replacing, then new ones can be made up to the original size and stitched on for two inches by the lower cheek buckle piece (see Fig. 6.2). You will not need to replace the complete cheek piece that passes round the winkers if the two buckle returns are in good condition. Simply cut the billet off from the end of the stitching and skive the remaining leather where the new billet is to be stitched. This will mean cutting through some stitches whilst skiving, but do not worry, you will restitch them as you replace the billet.

Fig. 6.2 The billets of a bridle, whether used for riding or driving, take a great deal of stress where the bit is attached. Always check here for signs of wear.

Winker stays

Considering the fact that winker stays are employed to support the winkers (blinkers) most of them are made from very fine pieces of leather, and will, in time, need repairing. The point of wear is usually along the stitch holes which hold the end of the stay in between the two pieces of leather that make up the winker. A regular inspection will reveal any weakness before the leather actually breaks. The repair requires the stitches holding the stay to the winker to be cut and the end of the stay trimmed, reinserted into the winker, and then restitched. You will of course be shortening the length of the winker stay by approximately ⅜ in which should not be noticed in use, or, the winker stay point at the centre of the head piece can be let out one hole to compensate (see Figs. 6.3 a and b).

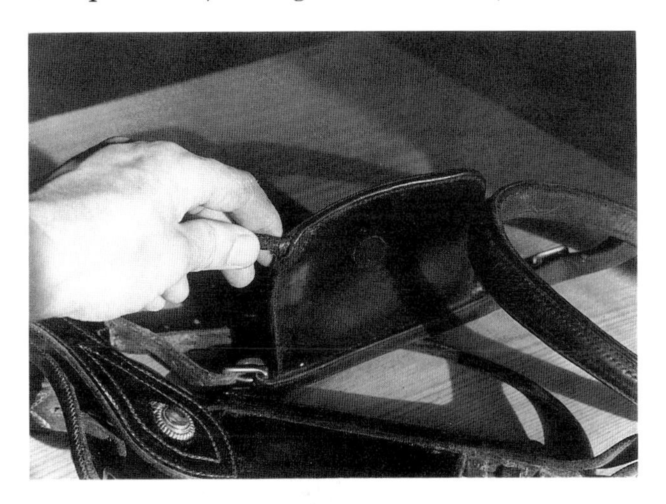

Fig. 6.3 a and b The end of the winker stay on a driving bridle will often wear at the point where it is stitched to the winker.

Breastcollar and Breeching Tugs, or Buckle Pieces

These narrow buckle pieces buckle onto the wither part of a breastcollar and are held by only six or seven stitches, and because they carry quite a considerable amount of weight, will wear in time

Fig. 6.4 A tug or buckle piece from a driving breast collar or breeching is shown here ready for stitching complete with its leather loops, buckle, D and filler piece.

(see Fig. 6.4). The easiest way to repair these is to patch the worn part of either the top or the bottom of the tug. A far better way, however, and one which requires only marginally more skill and leather, is to make a replacement part.

Carefully cut the stitches of the worn tug and lay the old piece of leather out on the bench to use as a pattern. Take all your measurements from this old tug, and if need be, cut new loops. If, however, the loops are in good condition, re-use them, not just to save leather, but to keep an original look to the set of harness. In use the new leather will not be seen, but the loops will (see Fig. 6.5).

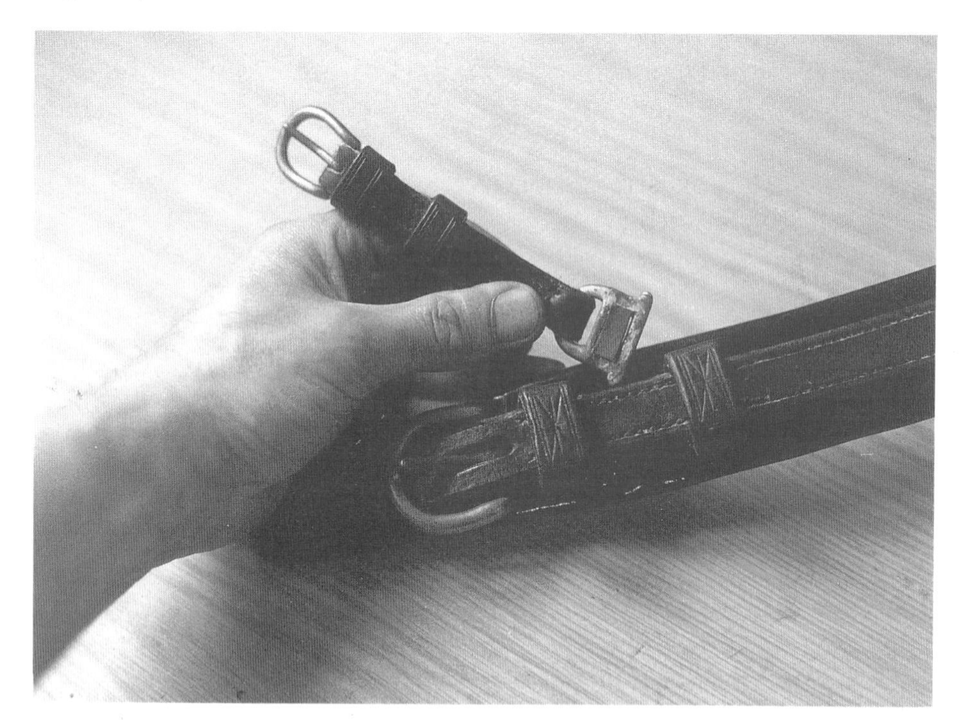

Fig. 6.5 The tug (buckle piece) that buckles onto the wither part on a breast collar is held by only six or seven stitches and will often require repair. When restitching it is best to always cut a new piece of leather to go inside the breeching D. A similar fitting is also used in a full breeching.

Traces

I strongly disagree with splicing traces. If they have broken, or, indeed, are about to break, replace them with a new pair. This may sound extravagant, but is not nearly so expensive as the cost of repairing your vehicle if the traces should snap in use, not to mention the poor unfortunate horse or pony.

I have on a few occasions reinforced a hole by stitching another piece of leather behind it. This may keep an exercise set going for a while but it is not ideal. The worn out trace could, however, be polished, rolled up tight and placed in your spares case for shows, although I hope that you never have to use it.

Shaft Tugs

The stitching on the tugs will wear if excess weight is put upon them because a vehicle is badly balanced, or if the screws of the tug stops on the vehicle's shafts are protruding. If the stitches do need restitching, be warned this can be a tough old job. The tug is made up of three or four thicknesses of leather, and, because it does not need to flex in use, can become incredibly hard and solid.

Use the existing stitch holes and your largest awl and two needles threaded with 5-cord or double thickness 3-cord. Be very careful to push the awl in straight, otherwise it is all to easy to snap a blade (see Fig. 6.6).

Fig. 6.6 Tug stitching will usually wear at the bottom where the shafts rest. Take great care when restitching not to use the awl at an angle or it will snap. Great strength may be needed to stitch through three or four thicknesses of old, dried out leather.

When you have restitched the tug, using the old stitch holes as a guide, dip it into water and flatten the stitches with either a small hammer, bone or wooden handle. This will help to reduce future wear (see Fig. 6.7).

Fig. 6.7 When you have finished stitching the tugs, flattening the stitches with a hammer or a bone will reduce future wear.

Pad

The lining of the pad can be replaced or to be more exact, over-lined, similarly to a riding saddle panel (see Chapter 4).

Although they do not wear as quickly as girth straps on a saddle, the points at the end of a pad will eventually need replacing. The original may well have been lined and swelled, i.e. the point was made up of two thicknesses of leather with a narrow filler inserted between the two to produce a rounded effect. However, a solid one-thickness piece of good quality leather will be adequate as a replacement. There are several ways of fixing the points onto the end of the pad flap. You may be able to stitch the replacement directly to the top part of the old point if it is still in good condition where it joins to the pad flaps. Otherwise you will need to cut the stitches around the end of the flap to release the backing piece of leather onto which the point is first stitched. The backing piece, together with the new point is then stitched around the end of the flaps. With some pads the point has been stitched directly onto the flap either on top or underneath, and if this is the case you will have to undo a couple of stitches or wire loops that hold the pad to the flaps (see Fig. 6.8).

Fig. 6.8 There are three main ways in which to attach a girth strap to a driving pad flap.

1 The strap is first stitched to a backing piece of leather which is then stitched to the underneath of the flap.
2 The strap can be stitched directly on top of the flap.
3 The strap is stitched directly to the underneath of the flap following the existing lines of stitching.

Traditionally, the lining of a driving pad was held onto the flaps by copper wire twisted to form a stitch. A small amount of coloured flock was incorporated into this twist of wire which formed a line of spots around the edge of the panel.

Perhaps because I came into harness making through working on saddles and bridles, I have always stitched, not wired, the panel to the flaps in the style of lacing up the panel on a riding saddle. The pressure and stress put upon a saddle is far greater than that put on a driving pad, but I have always preferred to treat them the same. I have had old pads in for repair in which the flock in the panel has settled and the wire has been just slightly proud of the pad – with remnants of horse hair in the wire!

If a pad came in for repair on which the wire had broken then I would always stitch using a curved needle and awl because I have seen too many pads in which the padding has settled and allowed the ends of the copper wire to come into contact with the horse or pony. Occasionally a working pad may need repair and its condition prevents lacing because the leather is in such poor condition. With this type of pad I see nothing wrong in stitching the panel to the pad directly through the top surface of the flaps and skirts using a single straight needle and awl. This would not be acceptable on a set of show-quality harness because the large stitches that would be visible on the top of the pad flap would be unsightly. You can still keep the repair as neat as possible by marking the stitches at equal distances along the edge of the flaps and if possible using the old lines of stitching as a guide.

Neck Collars

A well-fitting neck collar is advantageous in getting the best work out of an animal; an ill-fitting one, however, is best replaced with a breast collar. Bearing this in mind, it does, therefore, pay to look after a sound, well-fitting collar whether for work or show.

Only the very best condition will do for the show ring, so any re-covering is best left to an experienced harness maker, but with a collar used either for farm work, exercise or rallies there is no reason why you should not attempt to re-cover the lining. The most important thing to remember is that the straw of the body must generally be in good condition. Any slight unevenness of the part coming into contact with the animal can easily be corrected by covering it with a thick wool blanket prior to re-covering.

The inner lining

The inside lining on an old working collar, and most certainly a farm-type collar, could be of heavy-duty woollen cloth. In fact, the old oatmeal-coloured cloth with a blue check design on it is called collar check, and, if you can get it, is marvellous stuff. Otherwise any good quality, even ex-War Department blankets could be used to overline the inside (see Fig. 6.9).

Fig. 6.9 Many an old working collar has been given a new lease of life by relining with a piece of heavy duty linen or a heavy blanketing. Although the material in this picture is white, for a working collar any colour will do; dark blue can look very smart.

If a leather lining is to be used, select a thin, pliable skin such as panel hide or a cheaper basil. It does not need to be a light brown colour, a black skin will be acceptable. Leather will not stretch as easily as material, and may need dampening with a little water to make it more pliable when being worked.

Overlining a collar is a similar job to overlining a riding saddle (see Chapter 4). Cut the material to shape allowing ample turnover at all edges and stick it in place with a little Copydex-type glue. I have seen collars completely relined with just blanketing and glue – not a stitch in sight!

To effect a more workmanlike repair, however, start stitching along the length of the wale – the tubular-like hoop that runs round the front of the collar. You will find it is relatively easy to push the awl through the collar between the wale and the body of the collar; remember to turn the edge of the lining underneath as you go. Use large stitches and either pop stitch – making a large stitch in the outside groove of the wale and body and a smaller stitch on the surface of the lining – or lace using a straight awl and needle. Pop stitching looks far neater with a material lining than with a leather one (see Fig. 6.10).

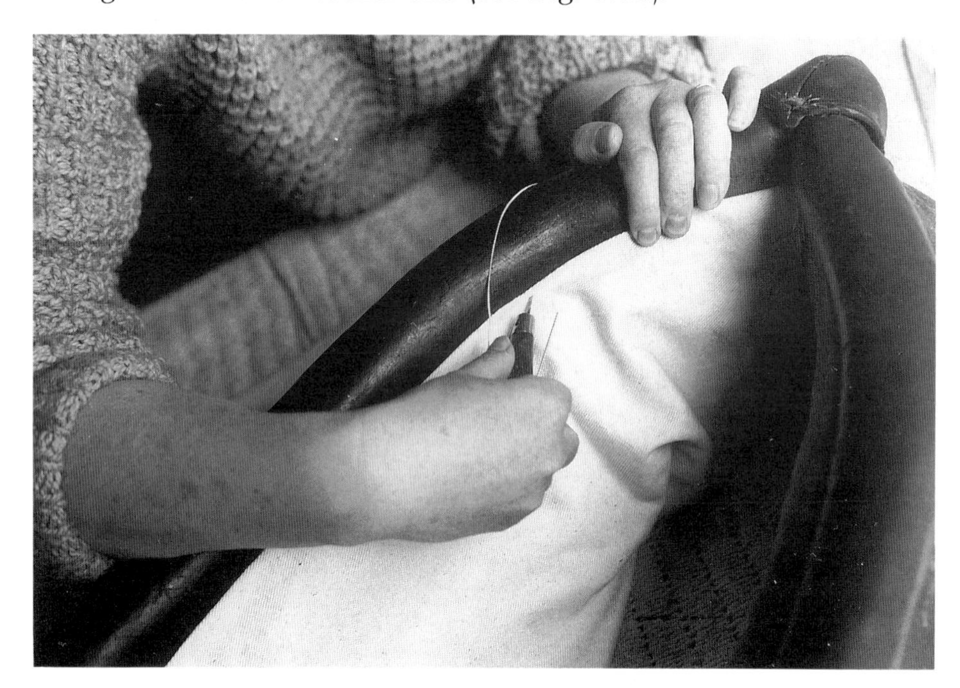

Fig. 6.10 Here the lining is being stitched onto a farm collar with the stitching going through between the round wale and the body of the collar. You will need a large awl and, again I stress, do not use the blade as a lever.

With one side of the lining stitched in place, pull it firmly over the body of the collar to the outer covering. Here it can either be stitched directly to the outside covering, or, if this is to be replaced also, to the existing inner lining and padding. As you are re-covering, you can leave the old padding in place providing it is sound. In both cases use a curved awl and needle.

The outer cover

If the outer covering is in need of repair it can either be patched solely on the points of wear – usually where the hame tugs come into contact with it – or removed and re-covered.

Fig. 6.11 Wear on a driving collar will often occur on the edge of the outer covering as shown here. The part that comes into contact with the trace clips of the hames, which can be seen just below the stitched repair in this picture, is particularly at risk.

To re-cover it completely, again choose a thin pliable skin or black patent-like plastic and glue this directly to the old covering you have removed. Turn the outside edge over and stitch. If you do not have a heavy-duty sewing machine this will be a long job.

Hames

The hames are the long metal, or wood and metal, pieces that run along each side of the collar and transmit the weight of the draft load to the neck collar.

Electroplated hames will always chip in use and often start to rust. These are not to be confused with close-plated hames where the brass is welded onto the steel hames and will last for years.

If your hames are past their best and starting to go rusty, there are three options.

1 Strip the tug buckles from them and have them replated. This is a long-winded job requiring much skill in reriveting and restitching, but undoubtedly gives the best-looking effect. For show use this is the only way.
2 Painting the hames with gold or silver paint. This may seem odd but my cob's farm collar has gold-painted hames and it does not look bad at all.
3 Covering the hames with a thin black hide, with just the loops for the hame straps, the rein rings and the buckle tugs protruding through. Cut the leather the length of the hame and the exact width to fold around the thickness of the steel. Dampen the leather and then pull it around the width of the hame, stitching it edge to edge on the underneath so that in use the stitches will not be seen. Use a curved needle and do not make the stitches too small. This type of repair, neatly carried out, can look very smart indeed. This is most probably the best option for working harness. Indeed many of the hames on the everyday exercise and working harness in the Royal Mews at Buckingham Palace have been covered in this way.

Webbing Harness

Webbing driving harness has gained popularity, not only for breaking an animal to harness, but also for exercising, rally drives and competition. Its two great advantages are:

1 cost;

2 maintenance.

Because it is often thought of as 'maintenance free', worn parts often go undetected until they eventually break. Most of the repair work on webbing harness involves replacing worn, fraying web, or renewing rusted or broken buckles. This is normally straightforward because any replacement part needs only to be cut to the pattern and stitched in place. I say 'only', but this actually disguises the fact that, if you do not have a heavy-duty sewing machine, stitching web by hand is a slow, but perfectly feasible, job.

Certain makes of web harness employ metal eyelets into which the buckle tongue fits. When these pull out you will need replacement eyelets and the right tools to secure them. These can often be bought from an agricultural merchant or, failing that, a tent and tarpaulin manufacturer.

Many sets of webbing driving harness have clips where the traces attach to the breast collar to ensure a quick release in an emergency. It is imperative that these clips are oiled regularly to keep them working freely.

My own set of web harness has been in regular use for three years now and has to date not needed any replacement parts. I have, however, checked it over regularly, and sealed any fraying nylon with a flame as soon as it showed the least sign of wear.

7

DRIVING VEHICLES

In addition to the horse's harness, the driving vehicle's leather work is also within the scope of your repair skills.

Shafts

On a traditional-style vehicle, shafts are normally covered with a thin hide to just past the tug stops and again to where the breeching Ds are attached. The friction of normal use and the added friction of transporting the vehicle with the shafts on shaft stands, may cause the stitching underneath to fray and require attention (see Fig. 7.1). Use a curved needle and oversew the edges together. If the edges of the leather do not meet, dampen the leather with water to make it slightly pliable (see Fig. 7.2).

Wings and Dashboard

These will not normally need repair, but when they do require attention it is often a complete re-covering job. This is within the scope of someone who has mastered the use of an awl and two-needle stitching.

If you are covering a dashboard or wings, it is best to clean the metal work with a wire brush once the old leather is removed, and paint with a rust preventer. Any top coat gloss paint is best left till after you have finished re-covering the metal frame.

Fig. 7.1 The leather at the ends of the shafts is in urgent need of repair. In practice the leather should not be allowed to get into such a poor state before stitching. When travelling vehicles, it helps to wrap the shafts with plenty of padding where they make contact with the shaft stands, and to strap them down securely.

Fig. 7.2 To restitch, use a curved needle and heavy thread. If the edges do not meet underneath, try soaking the leather with warm water first.

Remove the wings and dash from the vehicle then remove the old leather and discard it. This is one of the few jobs where it is best not to use the old part as a pattern for the new. Instead, cut your new covering – patent, plastic or bag hide, it really depends on the vehicle, money and materials available – approximately ½ in wider all round than the metal frame. Cut a second piece the same size and glue one onto a piece of woollen blanket-type material. This will give a slightly padded look and feel when the job is finished.

Mark the stitches all round at five, six or seven to the inch and again inside the line of the metal frame. If there are any metal supports within the outer framework of the wings and dashboard, these must also have stitches marked each side of them. Only one side of the new covering need be marked for stitching.

With a rubber-base glue such as Copydex, join the two pieces of the new covering together, with the exception of the outer two inches or so, into which the frame will fit. By gluing the two sides of the new covering together, you will eliminate the chance of them moving while you are stitching and it will also ensure that they lie flat together and will not bubble out on one side. If there are metal supports inside the framework, this is slightly more difficult because you will need to glue the pieces directly onto the framework. It may help to put a tacking stitch on the outside edge of each side of the work to hold it to the frame. These can be removed later.

Start by stitching around the outside of the frame using two needles and pull the stitches very firmly. This should pull the covering taut over the metal frame. When you have completed the stitching around the outside edge of the frame, start on the inside line of stitching, and, again, pull the stitches firmly. On some parts of the work you may be able to hold the frame in the clamps, but be careful not to mark the leather, on others it may be easier to stitch it while it is lying flat on the bench. I find this one of the most rewarding jobs I have had to do, because you can see the shape taking place as you pull on the inside line of stitches (see Fig. 7.3).

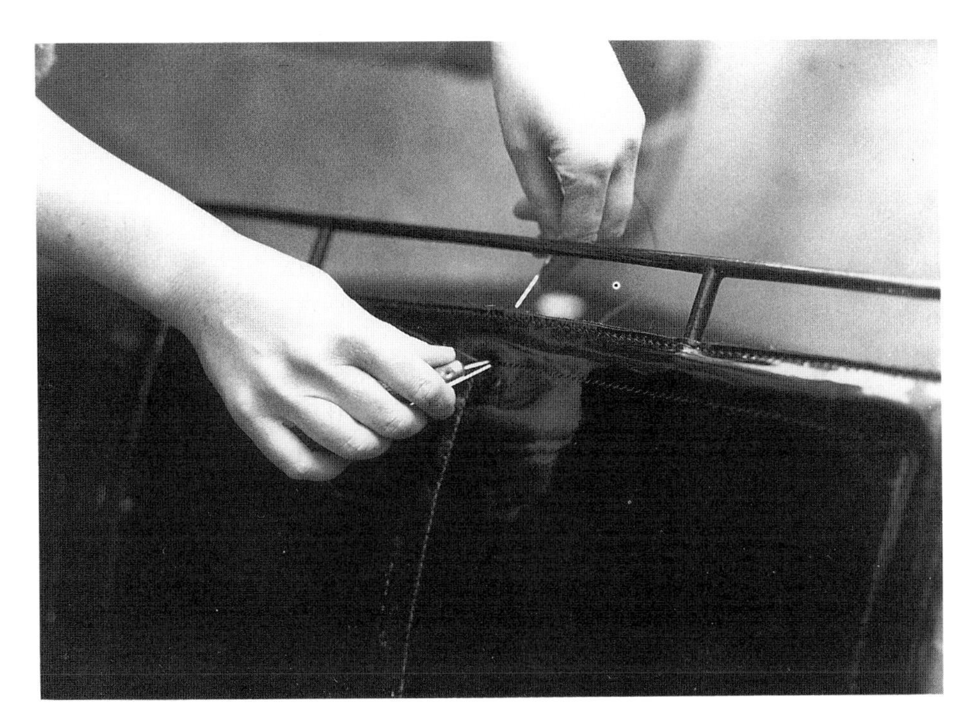

Fig. 7.3 Recovering a dashboard and wings is well within your capabilities once you have mastered the art of stitching with two needles. Start by stitching around the outside, then, once you start stitching on the inside line of stitches, you will see and feel the tension in the leather.

8
OTHER JOBS

Now that we have covered most aspects of cutting, preparing, marking out and stitching, there are an assortment of other jobs that can easily be tackled. Some of these are not confined to saddlery and harness.

Whips

Riding whips

The most common riding whip repair is to replace the leather loop at the end of the whip.

If you do not have the old loop as a pattern, cut a new one out of light hide approximately 8 inches long and 1¼ inches wide. Fold it in half and taper the ends to half the circumference of the whip end. Skive the two ends down for a neat repair. Now soak the loop in warm water and start binding it onto the whip using the heaviest cord you have to hand. By soaking the leather you will be able to pull the thread slightly into the leather, so making a stronger repair once the loop has dried out.

Hunting whips

There are four common hunting whip repairs.

1 The handle. This is easily repaired by using one of the modern superglues to secure the hook back onto the stock.
2 The keeper or loop at the end of the stock. If this has merely become detached owing to a worn binding, rebind the original

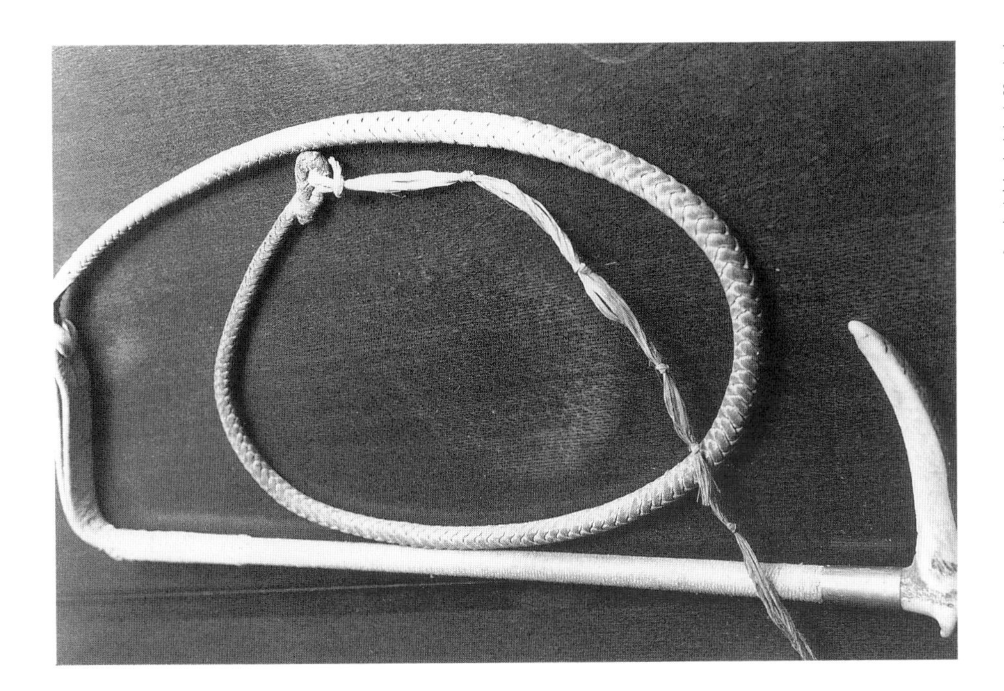

Fig. 8.1 A hunting whip showing hook, stock, keeper or loop, thong and lash. Note that the loop folds over towards the hook for ease of carrying and use.

keeper with heavy thread. As with a riding whip loop, first soak the leather in warm water. If the leather has broken and needs replacing, cut the new loop or keeper to the original pattern and bind on as above.

Many hunting whips have double thickness keepers that are stitched around the outside with an oval hole punched out to take the thong. If necessary these can be replaced with a single loop of strong leather cut approximately 12 inches long and one inch wide. See Fig. 8.1 showing a hunt servant's whip with just such an end. An important point to remember when binding a loop or keeper to a hunting whip is that the loop must face the direction of the hook of the handle. Look at Fig. 8.1 again and you will see what I mean. This allows the thong to be gathered with ease, held comfortably and is easier to crack when in use.

3 The end of the thong. This can easily break off, especially if it catches under a horse's hoof.

Dip the end in water, fold it over to form a loop and bind it round with a length of thick thread (see Figs. 8.2 a and b). This repair is actually stronger than the original plaited loop. I once repaired the master's whip like this out hunting while still riding my horse. I dragged the end in a puddle and bound it with spare hay string. It is still in use years later. One hunt servant I know always cuts the end off a new whip immediately he gets it and rebuilds the loop with cord. If you are at home in the workshop and not doing emergency repairs in the hunting field, a couple of stitches, when you have finished binding, will keep the end of the thread in place.

Fig. 8.2a Before stitching a new loop, thoroughly soak the end of the thong in water, thus making it pliable.

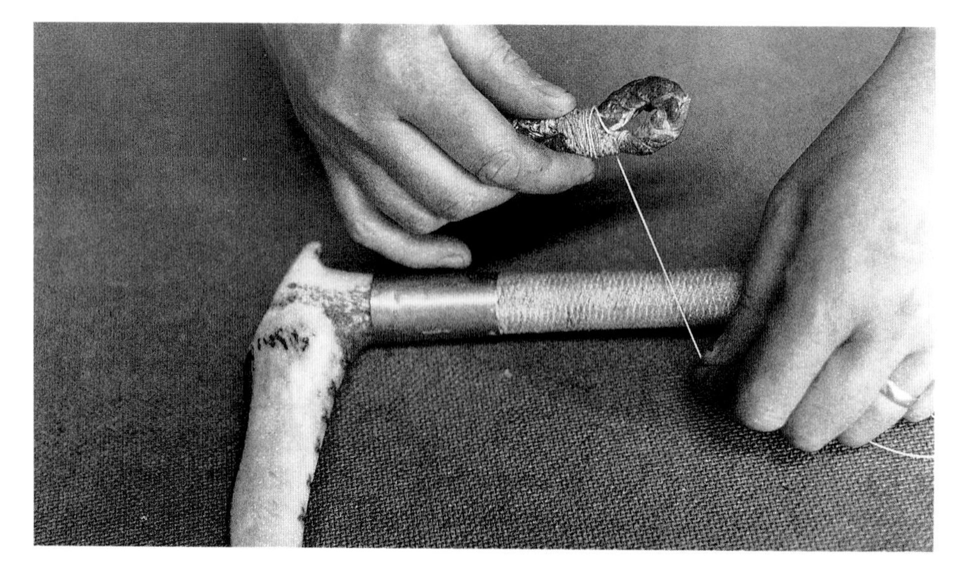

Fig. 8.2b When pliable, fold the end of the thong into a loop and bind together with the thickest thread to hand. When you have finished, a couple of stitches will keep the end of the thread in place.

4 The short lash on the end of the thong. The original is made from cotton cord but most hunting folk I know much prefer nylon hay string – it cracks much better.

Take a piece of hay string approximately two feet long, fold it in half and with one of the pieces of string, tie a series of knots around the other piece. This is then looped through the end of the thong (see Fig. 8.3).

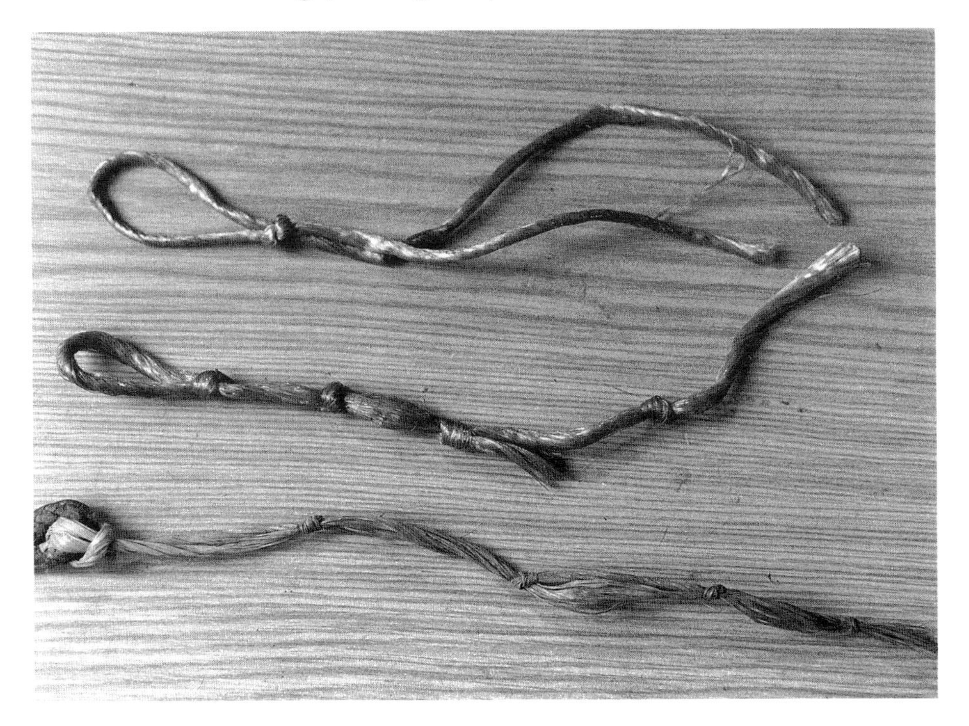

Fig. 8.3 This picture shows the three stages of making a new lash for a hunting whip using nylon hay string. Basically, once it is folded in half, one piece is knotted around the other. A driving whiplash can be made from three pieces of finer cord and after plaiting for a length, finish off with knots in the same fashion.

Driving whips

The traditional bow top driving whip uses a goose quill to attach the thong onto the stock. It is this delicate quill that gives the whip the elegant curve at its top – hence bow top – and the quill must be looked after when in storage. A whip reel is the best way to store a driving whip instead of leaving it propped up against a wall (see Fig. 8.4).

Great care must be taken with a whip of this type because, once the goose quill is broken, it *normally* needs a completely new thong and quill to look as good, and work as well, as before.

Fig. 8.4 The traditional bow top driving whip stored on a whip reel to preserve the delicate goose quill.

If, however, the damage is where the quill joins the stock, a repair can be effected by carefully removing the quill and thong, and rebinding, making sure that the damaged part of the quill is now overlapping the end of the stock. The previously damaged part is now bound with thread and varnished. Sometimes the damaged part of the quill is removed, if possible.

The end of the thong and lash may, from time to time, also need repair. Instructions 3 and 4 in the hunting whip section also apply to driving whips, but the repairs should be carried out much more delicately.

A replacement lash can be made from plaiting six pieces of stitching thread. Plait the lash from three strands (each of two pieces of thread) and, as you get towards the end of the plaiting, start to lose a piece from each strand. Do this by cutting off one of the pieces every two or three plaits so that the lash takes on a tapered look.

Boot Tops

Apart from improving the appearance of a pair of boots, be they leather or rubber, leather tops can also be fitted to lengthen a pair of boots. The top has a turned edge, and the end of the return is stitched to the top of the boot.

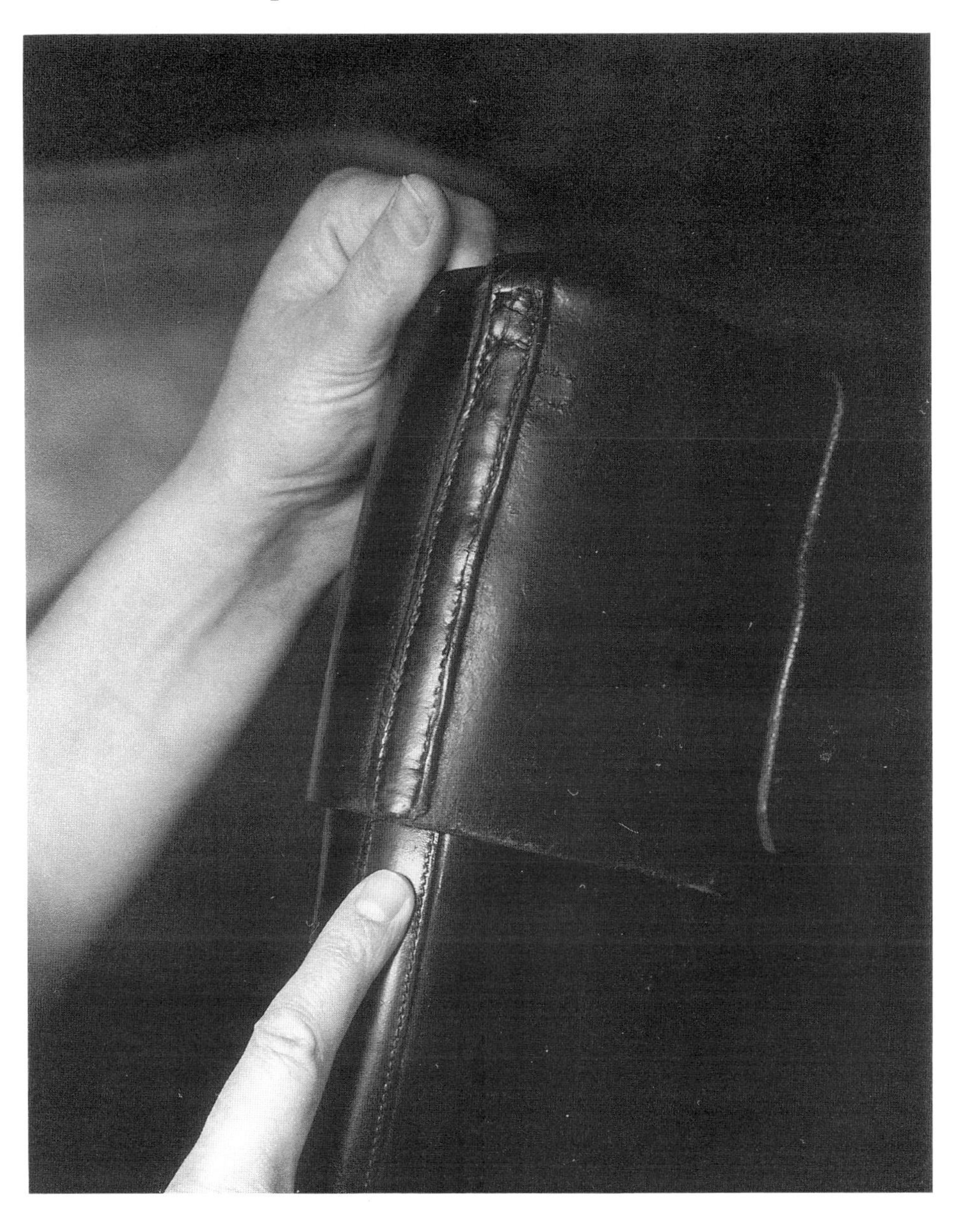

Fig. 8.5 Before stitching leather tops on a pair of boots, always check that the back seams of each match up exactly.

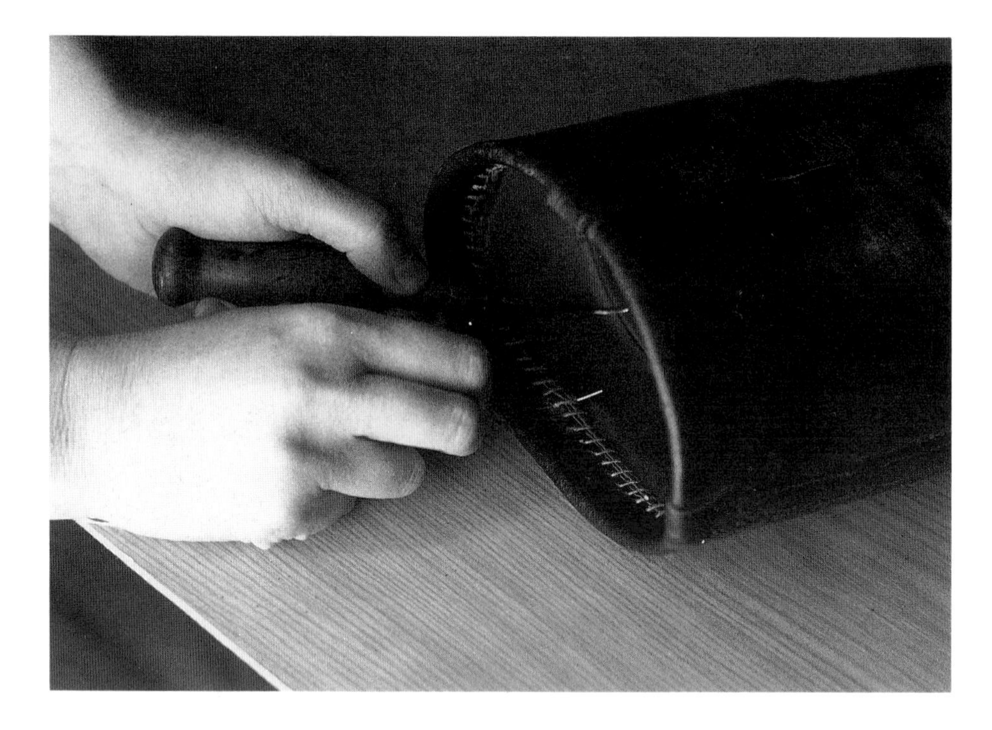

Fig. 8.6 When either restitching or replacing a pair of boot tops always use a curved awl and needle so that you can stitch from just one side of the work.

To restitch tops, or transfer a pair of tops from one pair of boots to another, start at the back, keeping the seam of the boot top in line with the back seam of the boot. Use a curved needle and awl so that you can stitch from just one side of the work. This way the stitches will lie inside the boot and not be seen. Finally, tap the stitching flat with a hammer, to reduce the friction to a rider's leg (see Figs. 8.5 and 8.6).

Briefcases and School Satchels

When I had a shop in Wales, every year around the end of August, I was inundated with school satchels needing repair ready for a new school term. Most of the satchels only needed restitching, mainly in the corners, and could easily have been carried out at home. The schools could even have included it in a craft lesson! To repair, simply restitch with a single needle using existing stitch holes; there is no need for an awl or a pair of clamps.

Old Style Leather Camera Cases

Fig. 8.7 Although leather for camera cases has now given way to waterproof plastic, there are still many old cases around needing restitching. With old leather like this great care has to be taken when using the awl so as not to split the leather. Pull the stitches gently to avoid pulling them completely through the leather.

Although these repairs are similar to those for briefcases and satchels, greater care must be taken when stitching because the stitches are finer, and an awl will normally be needed. In Fig. 8.7 you will see that the awl passes through both pieces of leather at 45 degrees. Take care not to split the leather with the awl.

Two-needle stitching will ensure an even tension when pulling the stitches tight, but they should not be too tight; most old cases have had little polish and the leather is often very dry and weak. Too much pressure will pull the stitches completely through the leather.

Car Hoods and Tonneau Covers

Repairs to these items are not so frequent now that small, open-top sports cars are out of fashion, but I have repaired several hoods including my own. It is a lot easier to restitch a seam or window by hand with the hood still attached to the car than to remove it to go on a machine, but if the whole hood or cover has to be restitched, it would be easier to remove it and use a sewing machine. Use the heaviest thread you have, or double thickness plaiting thread, and keep it well waxed while stitching. Beeswax can also be rubbed along the finished stitching to help waterproof it.

Bellows

It is to be hoped that the leather work will not be so far gone that you cannot remove it and use it as the basis for a pattern.

With a nail claw, carefully remove all the large-headed upholstery nails around the edge and above the metal tube. Cut a new piece of leather to shape using a soft but strong skin such as lining hide. Before starting to nail on the new piece of leather, check the condition of the valve on the inside of one of the wooden sides of the bellows. This valve allows air to enter the

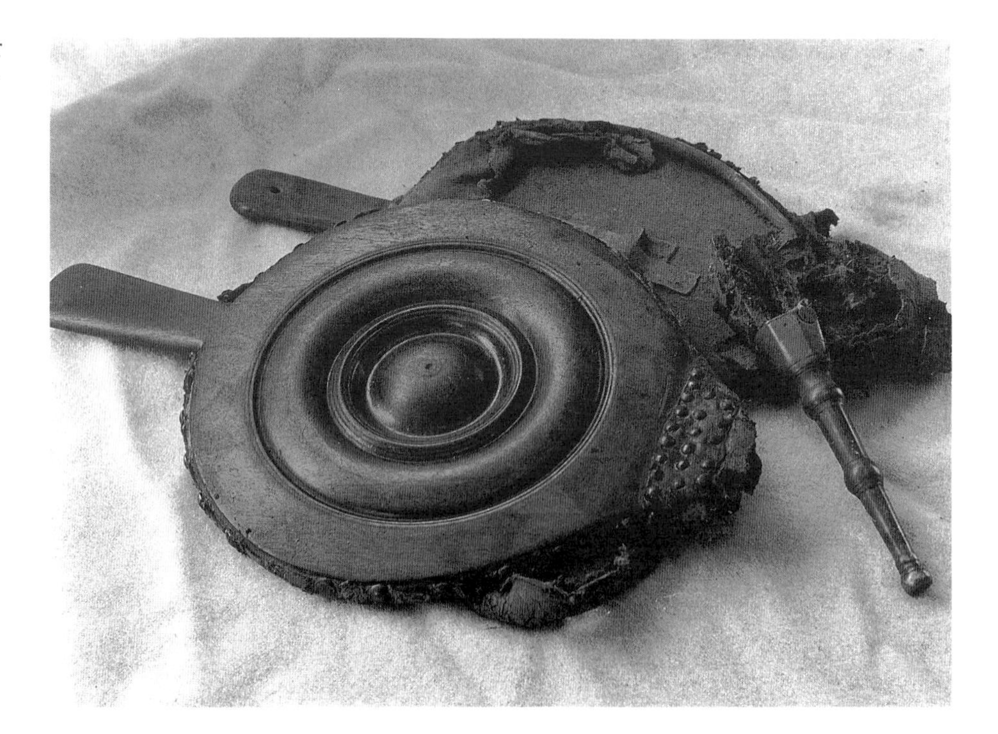

Fig. 8.8 A beautiful pair of bellows requiring new leather around the sides. Jobs like this will often come into the saddler's work shop and make a pleasant change from working on saddles and bridles.

bellows, then closes to ensure all expelled air passes through the metal tube or nozzle. This valve is simply a block of wood nailed to the inside of the bellows with a stout piece of leather to act as a hinge. Once the main leather part is nailed on, the triangular lower piece of leather that connects the wooden sides to the brass nozzle can then be nailed in place.

Try to use the old nails because they will have acquired a nice antique finish, but, if they need replacing, you should be able to buy upholstery nails from a hardware store (See Fig. 8.8).

9

TIPS FOR AVOIDING THE NEED FOR REPAIRS

There are numerous tips for the preservation of saddlery and equipment, many of which you will know. In this chapter I reiterate some of these tips which are listed for easy reference.

Saddles

Stirrup leathers

Keep the stitching well oiled because this reduces friction. If you need to restitch them, consider shortening the leather by four inches so altering the points of wear caused by stirrup bar and stirrup iron.

Three-fold girth

Check there is a piece of removable material inside the girth and take this out regularly to soak it in oil. If there is no material insert, cut a piece of blanket-type material 2½–3 inches wide, and the length of the girth, and place this inside the girth (see Fig. 9.1).

Saddle seat

Never put a saddle down with the cantle coming into contact with stone, brick or any rough surface. It is better to wrap the girth over the top of the cantle so it acts as protection (see Figs. 9.2, 9.3 and 9.4).

Fig. 9.1 Always keep the inside of a three-fold girth supple by oiling the removable piece of material. This can be made from either a piece of woollen blanketing, or heavy cotton or linen. It is an easy job to cut a piece of material to go inside if required.

Fig. 9.2 Once the seat has split like this, the only viable repair is to patch over the split which can look unsightly. It is possible for a good saddler to reseat the saddle but this involves stripping the seat down to the tree webs and is seldom an economic proposition.

Fig. 9.3 The seat of the saddle is very much thinner than the skirt and flaps so great care must be taken to avoid damage. All too often a saddle is rested against a rough stone or brick wall, which can cause serious damage to the cantle.

Fig. 9.4 A far better practice is to make a habit of always wrapping the girth, be it leather or material, over the cantle to give it protection.

Girth straps

Try to take the strain off the normal holes used by girthing up higher or lower on the offside when saddling up. When cleaning, make sure that the top of the girth straps, where they are stitched to the tree web, are well oiled on both sides if possible to avoid friction (see Fig. 9.5).

Skirts

Try not to pull the skirts upwards too hard, especially when adjusting the length of your stirrups while mounted, because this weakens and eventually breaks the stitching. If the silver saddle nail in the front pulls out, or a hole is worn in the leather around it so that the skirt can be pulled away from the nail, then it must be repaired immediately. Failure to repair this can lead to the whole front part of the skirt tearing away from the saddle seat. Once this has happened the only repair possible will look just what it is, a patch (see Fig. 9.6).

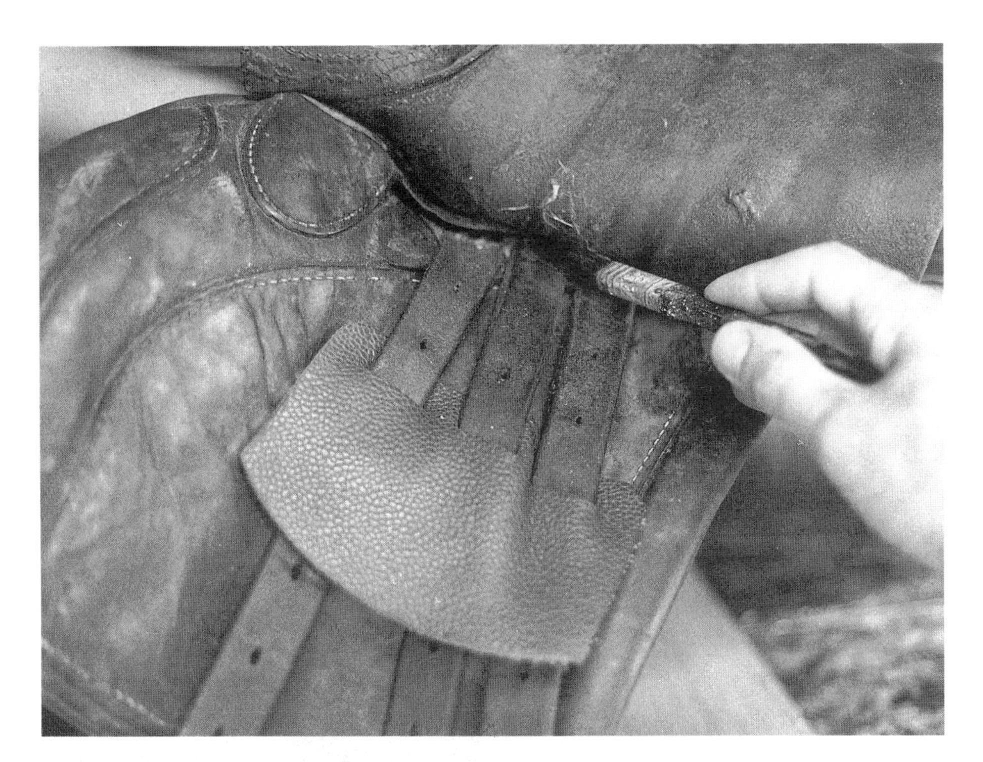

Fig. 9.5 When cleaning, always oil the stitches attaching the girth straps to the tree web. These stitches are longer than most and, even though they are made with the heaviest of thread, they can wear quickly if allowed to dry out.

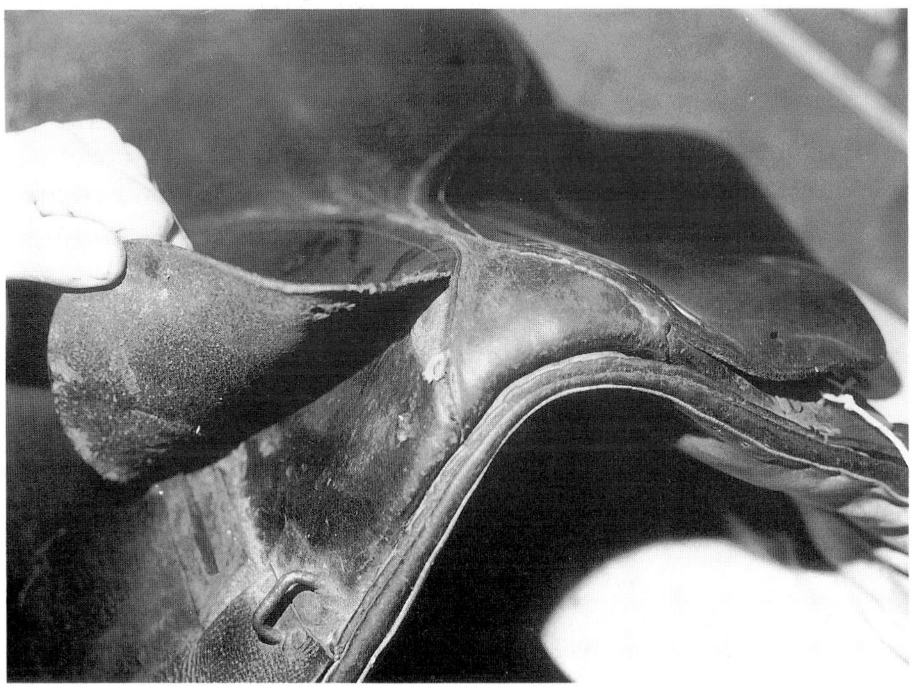

Fig. 9.6 The saddle nail holding the skirt in place has worn through the leather placing pressure on the seam with the seat when the skirt has been lifted to adjust the stirrup leather. With time this has caused the seam to split apart. To repair this saddle, a patch slightly larger than the saddle nail was placed between the nail and the skirt to act as a washer. The top of the skirt was then carefully stitched to the seat using small slanting stitches.

Flaps

Never carry a saddle by the flaps. It puts far more pressure on the stitching, tacks and saddle nails than they were designed for. One of the few repairs required on saddle flaps is the patching of the holes caused by the girth buckles wearing through from underneath. I know it can be argued by certain riders that buckle guards place another thickness of leather between man and beast so restricting the feel of the animal, but they do save wear on flaps (see Figs. 9.7a and b).

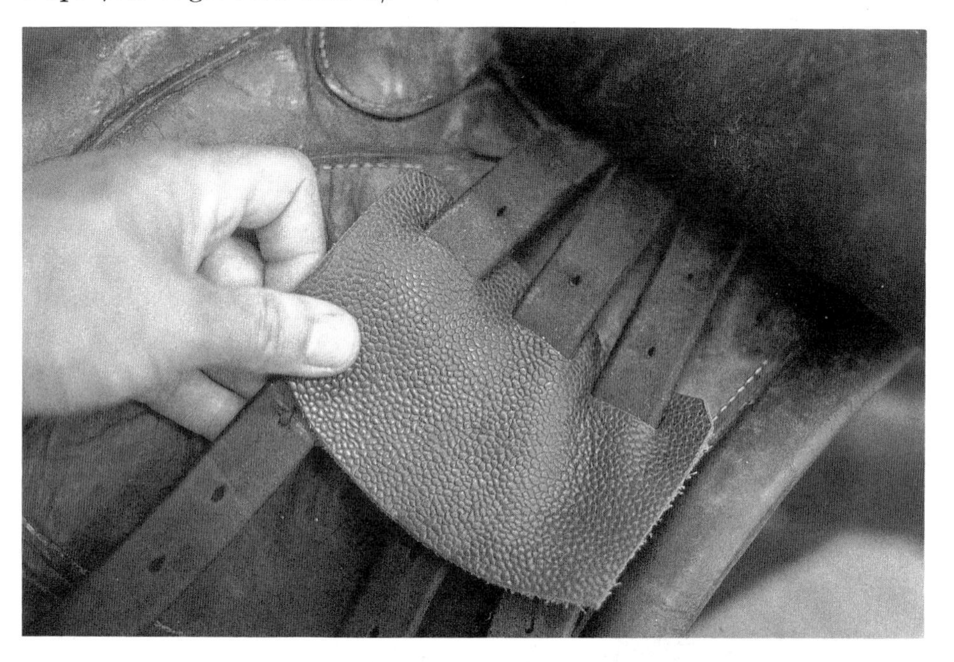

Fig. 9.7a A simple buckle guard can save the flaps considerable wear.

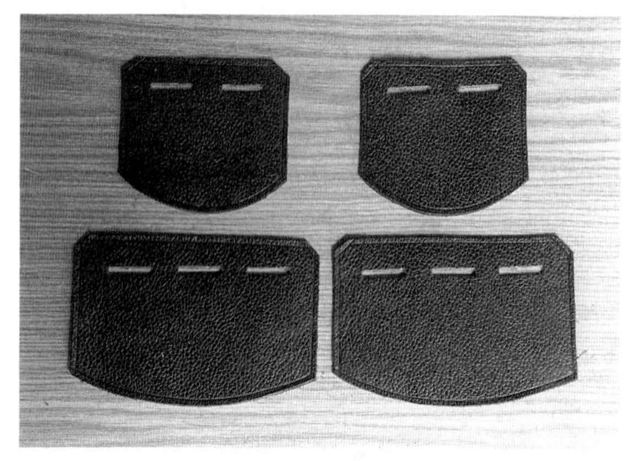

Fig. 9.7b A three strap buckle guard will measure approximately $5\frac{1}{2} \times 4$ in, and a two strap buckle guard will measure approximately 4×4 in.

Panel

When cleaning a saddle always gently fold the knee rolls outwards to clean and oil the inside folds of the leather (see Fig. 9.8). This is especially important if the outside lining is leather. If the

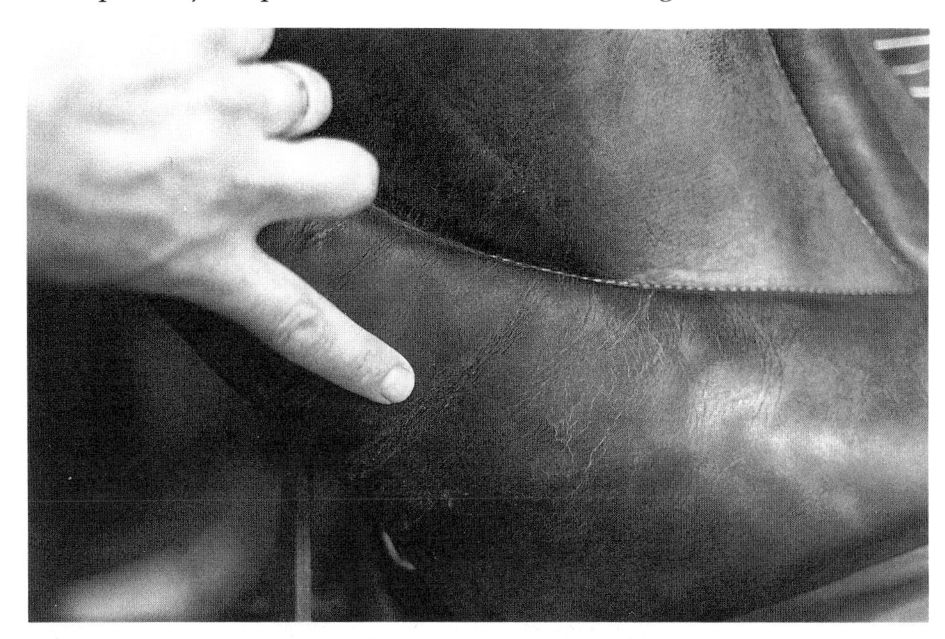

Fig. 9.8 The saddle panel can often be overlooked when cleaning and consequently will dry out, crack and eventually split. When cleaning, gently pull the panel outwards to make sure the oil feeds into the creases of the leather.

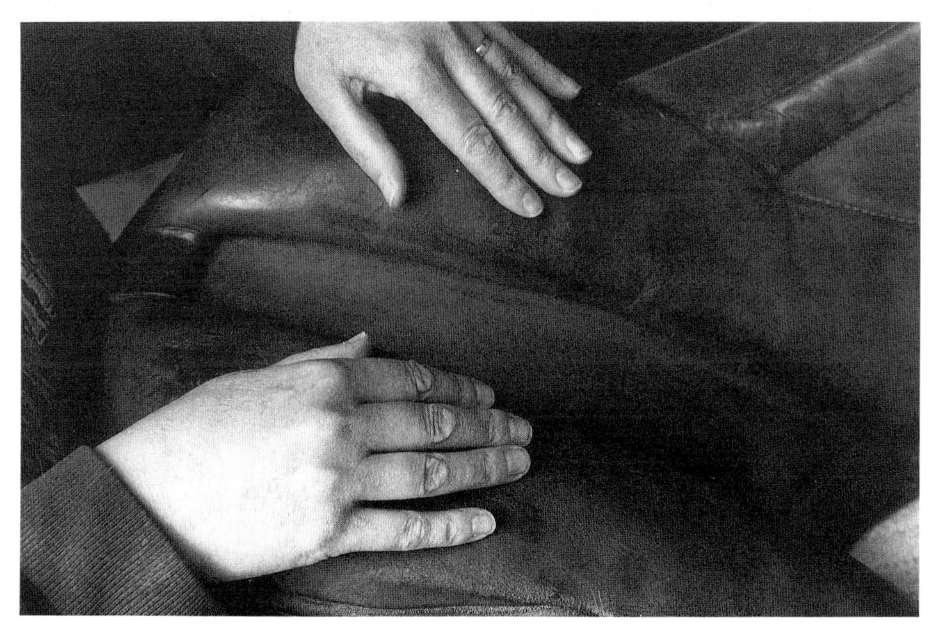

Fig. 9.9 The level and evenness of a panel can be judged by looking along its length and running your fingertips over the bearers. Any unevenness either in the size of the bearers or any lumps in the stuffing should be corrected by a good saddler.

lining is linen then this can be washed with warm soapy water. Old style surge panels can only be brushed with a good stiff dandy brush. Saddles with material-lined panels should always be used with a numnah because this soaks up the horse's sweat and stops it being absorbed into the flock stuffing of the saddle causing it to become hard, lumpy and impossible to restuff (see Fig. 9.10). Keep the stitching at the bottom of the panel well oiled especially the 3–4 inches where the girth passes over the stitches (see Fig. 9.11).

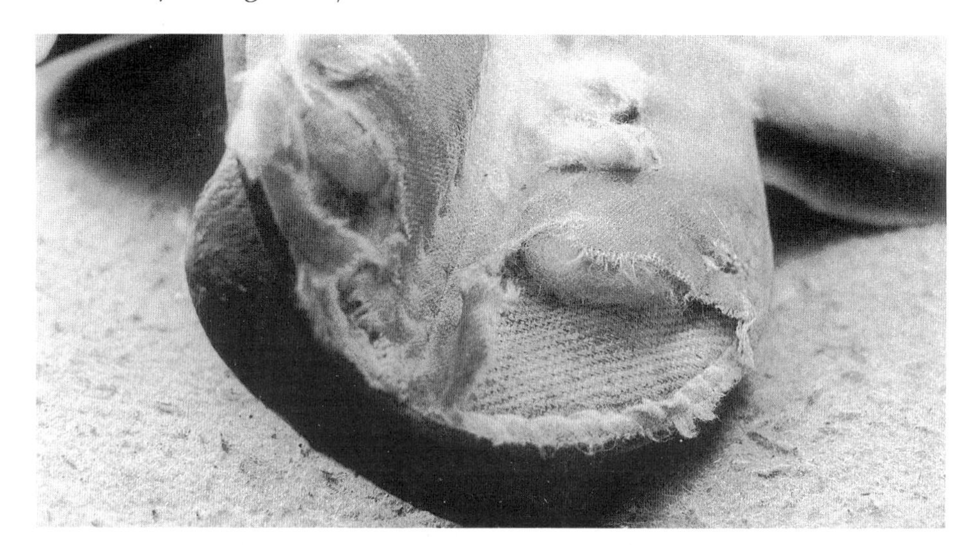

Fig. 9.10 An example of what can happen to an old-style serge-lined panel. The horse's sweat has soaked into the saddle flock and has rotted the material lining. A numnah will help protect the panel if it is washed regularly.

Fig. 9.11 The bottom part of the panel between my thumbs will always have heavy wear because this is where the girth lies. Again, extra oiling will help prolong the life of the stitches.

Breastcollar Ds

Considerable force is put on a breastcollar and this should be borne on the Ds or staples which pass through the tree and are secured underneath. Ds held on by thin strips of leather and nailed to the tree have insufficient strength; they are adequate for tying a riding coat to but not for attaching a breastcollar (see Figs 9.12 and 9.13).

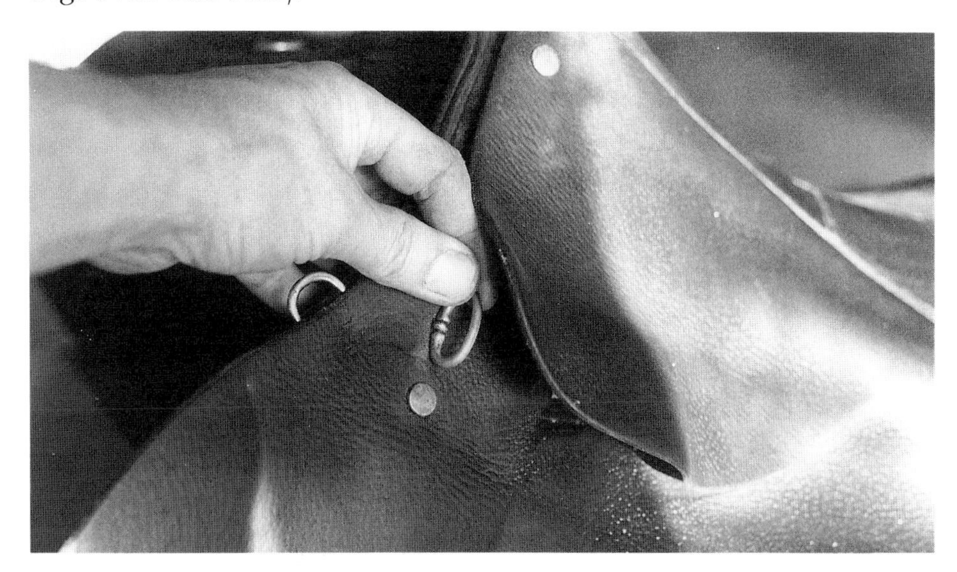

Fig. 9.12 If you are using a breastcollar then the D to which it should be attached is shown in this photograph. This folded down staple passes completely through the saddle tree and is then opened out underneath.

Fig. 9.13 This shows the split metal pin holding the D to the saddle as opposed to the other D in this photograph which is held in place by a thin piece of skin and a couple of small tacks. Its purpose is more ornamental than anything else because it has no real strength.

Bridles

Billet ends

Always keep the inside of the billet well oiled where it comes into contact with the bit.

Head piece

Make sure that there is a small hole at the top of the cheek pieces. This helps to prevent splitting. If there is not then insert one using the smallest hole of a wheel punch (see Figs 9.14 and 9.15).

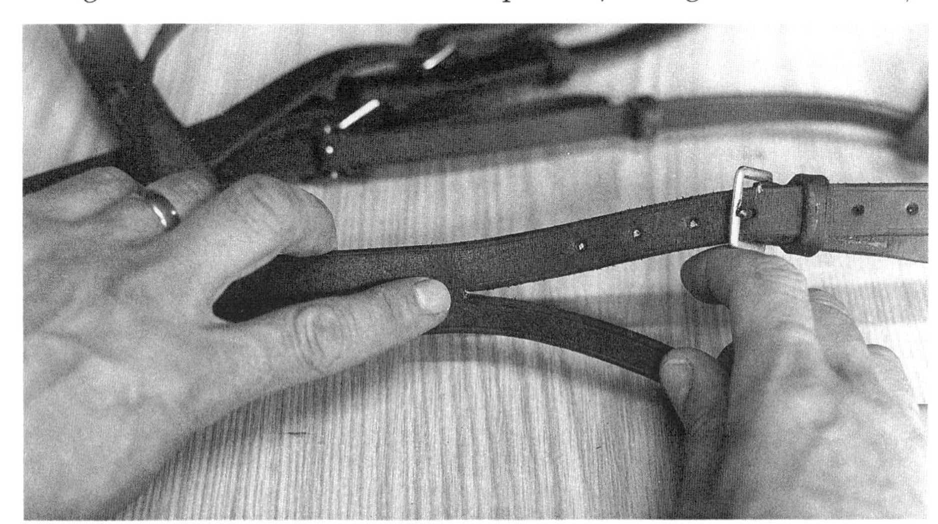

Fig. 9.14 A small hole at the start of the cheek and throat piece will help to prevent the split in the piece of leather continuing any higher in the bridle head piece.

Fig. 9.15 It is a very simple job to punch a hole at the start of the split if there is not one there already. If you have a leather balding girth, check that there is a hole at the end of every split in the leather.

Martingale and Breastcollar

Pay special attention to the loop that the girth passes through because this holds mud and sweat. Keep it washed and well oiled.

Numnahs

Always pull numnahs well up into the arch of the saddle before girthing up. This prevents them splitting along the centre seam and also helps to prevent the horse's wither being rubbed (see Fig. 9.16). Always wash a numnah regularly to rinse the horse's sweat out of the fibres, especially in hot weather.

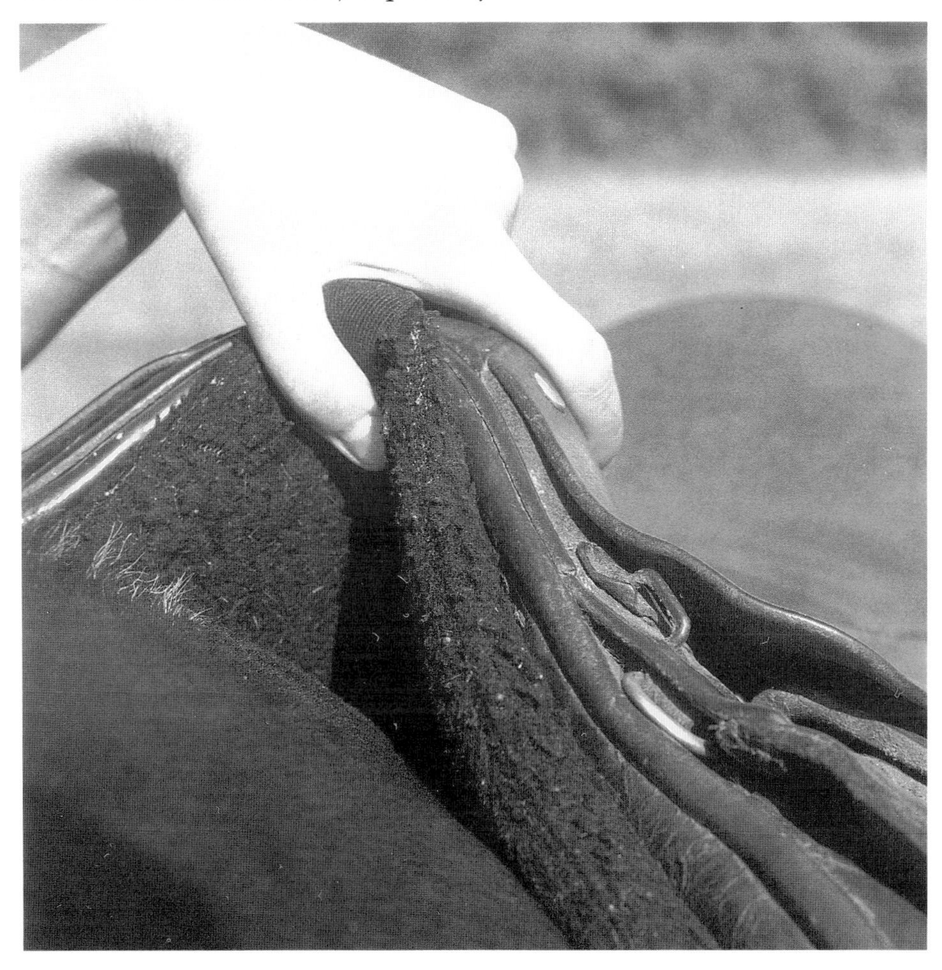

Fig. 9.16 Before girthing up always pull the numnah up to fit the saddle rather than letting it lay against the horse's wither. This not only helps to keep a passage of air along the spine and prevent rubbing but it will also stop the numnah splitting.

Buckles

When cleaning equipment, always dribble a little oil inside the fold of leather that turns around the buckle to stop the inside cracking (see Fig. 9.17).

Fig. 9.17 A little oil dribbled inside the fold of the leather will reduce friction, and therefore wear, between the leather and the buckle.

Headcollars

Where possible use a headcollar with a ring instead of a square at the centre of the jowls and backstay. It is far stronger because, when tying up, the stress is evenly distributed around the ring instead of the four corners of a square.

Nylon Webbing Headcollars and Harness

At the first sign of fraying, seal the webbing with a cigarette lighter, match or a gas flame (see Fig. 9.18).

Rugs

Anti-sweat

Always buy anti-sweat rugs one foot shorter or one to two sizes smaller than normal because, in use, these rugs settle down and

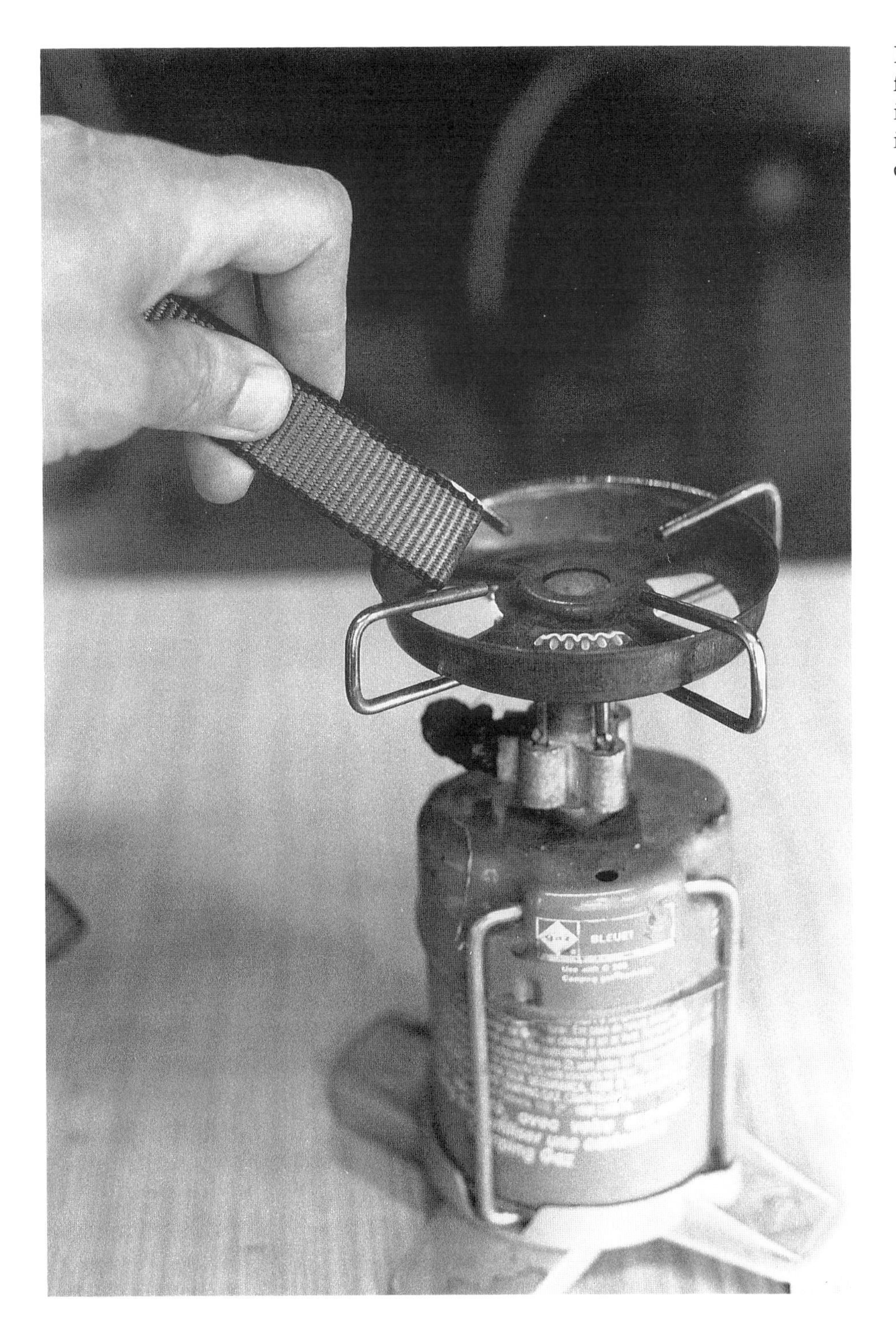

Fig. 9.18 Seal any fraying nylon webbing, be it a headcollar or harness, in a flame at the first sign of wear.

stretch along their length. A rug that fits correctly when newly put on soon drops down over the horses tail, and if the horse has a tendency to lean on his tail when travelling, the wear soon makes the holes even larger. Stitching up a string mesh rug once it has pulled apart is a devil of a fiddly job!

Wash the rug regularly to remove sweat. Check your horse box or trailer for any protruding nut or bolt heads which tend to catch and tear travelling and anti-sweat rugs if a horse brushes hard against them (see Fig. 5.6).

New Zealand

Contary to popular belief, chrome leather does get brittle and crack, so the fittings on New Zealand rugs must be oiled to keep them supple. Clips soon rust or get dirt jammed in them so keep them free moving and well oiled (see Fig. 9.19). Repair tears immediately because a small tear can quickly become a major job.

Fig. 9.19 By oiling all clips regularly they will remain free moving and therefore safer in use. New Zealand rug leg straps usually have clips at one end and these should constantly be cleaned and oiled especially when storing during the summer.

Driving Harness

Much of what I have written about saddles and bridles applies equally to driving harness with some additions.

Crupper docks

When they are not in use always keep crupper docks buckled up to maintain their shape.

Traces

Try to use the end slot of the traces when attaching a vehicle, so that if this should break you have another hole to get you home. You should always have at least two long slots at the vehicle end of the trace.

Webbing harness

Religiously seal any fraying ends and keep any clips well oiled and free moving. Wash regularly to remove sweat and mud. Most roko type buckles are not rust resistant, so if, when you take the harness off, they are wet, bring the harness inside to dry somewhere warm rather than letting them rust in a cold tackroom. A little oil dribbled inside the middle of the buckle will help to prevent rusting.

Driving Vehicles

Traditional shafts

Check that the screws attaching the tug stops to the shafts are the right size, flush, and will not cut into the tugs. Even leather tugs can quickly become cut up inside if used against tug stops with protruding screw heads (see Fig. 9.20).

All-metal shafts

Make sure that there are no odd spots of weld near where the breeching straps or tugs touch because they will eat through nylon webbing very quickly (see Fig. 9.21).

Storage

When storing an item of saddlery or harness for any length of time, and especially during the winter when the damp can attack metal work, always try to take each item apart so that buckle tongues do not stay buckled to leather work. Buckle tongues,

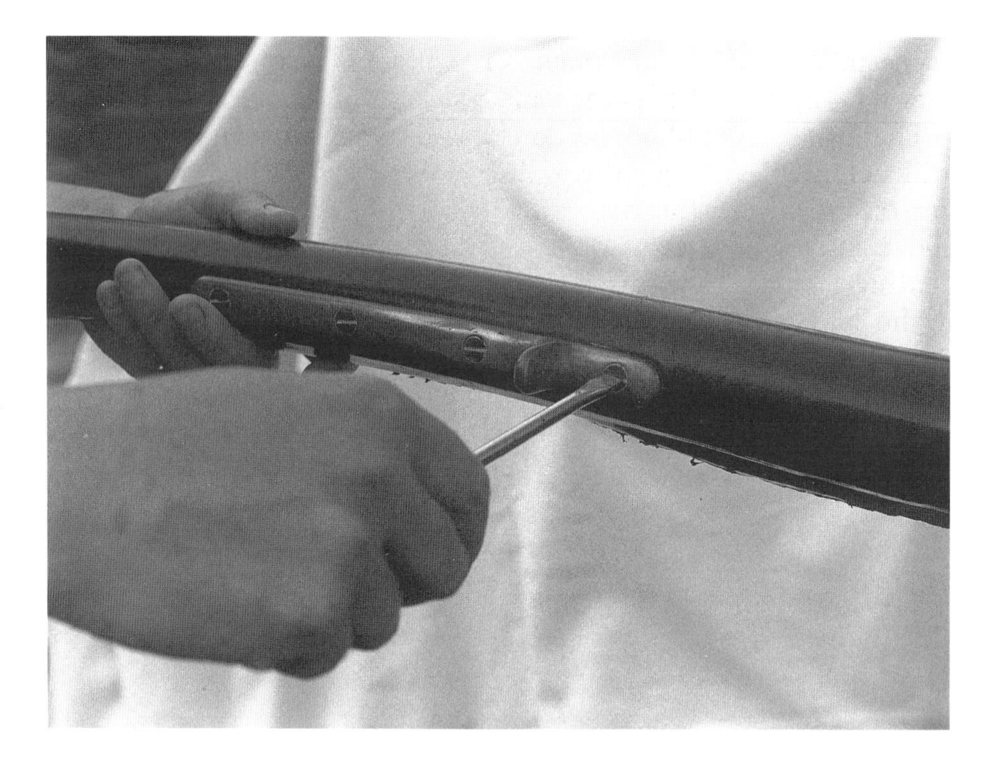

Fig. 9.20 Always check that the correct size screw has been used to secure tug stops. Any protruding screw head will very quickly cut into the inside of the leather tugs. It is also aesthetically correct to finish the screws with all slots in the heads running in the same direction as seen in this picture.

Fig. 9.21 Just the smallest piece of weld left on the shaft and coming into contact with leather or nylon-web tugs (more usual with an exercise cart) will do considerable damage. So check around the tug stops and breeching Ds and if there are any rough spots, use a file to smooth them off.

even stainless steel ones, have a nasty habit of rusting, and this eats its way into the leather and weakens it severely.

The leather should be well oiled, and wrapping it in newspaper will help keep it in good condition. What should never happen, but often does, is the hanging of a lovely set of leather harness above an open fire, virtually allowing it to cook to death. Central heating radiators are just as bad for leather. It is fine to do this if the harness has finished its working days and is purely for decoration, but to dry out a set in this way and then expect it to resume work after a quick oil could prove to be dangerous.

CONCLUSION

Over the years, apart from purely saddlery jobs, I have been asked to turn my hand to repairing harnesses for man-eating guard dogs; a harness for Caesar a dog who, after a throat operation, could not wear a collar round his neck; artificial arms and legs; car hoods and seats; hay binder canvases; veterinary horse gags; and, best of all, a harness used when hanging people on a film set. This harness went under the actors' costumes and supported their fall as they were 'hanged' with a noose round their neck. All in the best Hollywood tradition of course. I worked meticulously on that job, and checked my work again and again, and then once more!

So, as you can see, by first tackling simple jobs requiring only the most basic of tools and progressing to more complicated jobs, while adding to your tool collection, you will eventually be capable of handling a variety of ordinary, interesting, and sometimes most unusual, repairs.

APPENDIX

List of suppliers and Alternative Sources of Supply

Craftwares Ltd., Home Farm, Hotham, York, Y04 3UN. Tel. 0430 423636.
Good range of tools, webbing and buckles. Leather available to order.

Abbey Saddlery and Crafts Ltd., Marlborough Close, Parkgate Industrial Estate, Knutsford, Cheshire, WA16 8XN. Tel. 0565 650343.
Good range of tools, webbing, leather and buckles.

J. T. Batchelor Ltd., 9–10 Culford Mews, London, N1 4DZ. Tel. 071 254 2961/8521.
Suppliers of buckles, tools and leather. For personal callers there are leather bargains on the 'seconds' shelf.

Le Prevo Leather, Blackfriars, Stowell Street, Newcastle upon Tyne, NE1 4XN. Tel. 091 261 7648.
Suppliers of basic tools as well as buckles and leather. Helpful advice given to callers at the shop.

Tambour Supplies, 4A Penbeagle Ind. Est., St. Ives, Cornwall,

TR26 5ND. Tel. 0736 793305 (Factory) 0736 794411 Weekends/evenings.
Good range of canvas and webbing together with their fittings.

Bakers, Colyton, Devon.
Tanners of a wide range of leather.

Arthur Hart, Crewkerne, Somerset.
Weavers of narrow webbing. Bargains in 'seconds' and end of range materials for personal callers.

Hoopers Saddlers Shop, 52 Short Acre Street, Walsall, W. Mids., WS2 8HW. Tel. 0922 23639.
Complete range of leather, tools and fittings.

C. F. Osborne and Co., 125 Jersey Street, Harrison, New Jersey 07029, U.S.A.
Full range of leather tools.

U.K. agents for C. F. Osborne:
H. Webber and Sons Ltd., Teulon House, High Street, Ripley, Surrey, GU23 6AY.

In addition to the main suppliers previously listed, many of the items needed for repairs can be bought in small quantities in most towns.

Rings and clips	Hardware stores, ironmongers and farm suppliers.
Cotton and nylon webbing	Boat chandlers, hardware stores, ex-War Department stores.
Beeswax	Chemists, craft shops, health food shops.
Needles	Blunt-ended darning needles from needlework shops.
Thread	Sold as carpet thread, or plaiting thread when it is sold in camping and saddlery shops.

Canvas	Tent manufacturers and menders, old jeans and blankets.

I mentioned previously that all repairs in this book can successfully be carried out without a sewing machine. Many of you, however, might like to equip your workshop with one, so the following addresses could prove useful.

Sew Rite Sewing Machines, 45 Ropewalk Road, Llanelli, Dyfed, SA15 2AG. Tel. 0554 759833, Mobile 0836 659315.

J & B Sewing Machines Co. Ltd., 3 Coverack Road, Newport, Gwent. Tel. 0633 266495.

S.M.E.A. Co. Ltd., Unit 5, Heron Road Ind. Estate, Rumney, Cardiff. Tel. 0633 266495.

Other addresses can be found by looking through your local *Yellow Pages*, although many firms only deal in domestic sewing machines not industrial machines.

I have mentioned the auctioneering firm of Thimbleby and Shorland. This firm holds the very specialized horse-drawn vehicle and harness sales in Reading. They are held during, March, May, September and November each year, and provide an opportunity to buy both new and secondhand tools at auction. Many of my tools have been bought this way. In addition, at most sales there are leather, harness fittings and boxes of assorted old buckles on offer.

Apart from buying tools and materials, a day at one of their sales is a day well spent just looking at the sets of harness, driving accoutrements and of course the varied selection of vehicles.

Thimbleby and Shorland, 31 Great Knollys Street, Reading, Berkshire, RG1 7HU. Tel. 0734 508611.

BIBLIOGRAPHY

I recommend you read all the following books.

Saddlery by E. Hartley Edwards Pub. J. A. Allen
The first saddlery book I bought and most probably the best. A true encyclopaedia of equipment and its uses, written by a knowledgeable and highly capable horseman.

Saddlery and Harness Making by Paul Hasluck Pub. J. A. Allen
There is a huge source of knowledge in this book, but it is heavy reading because it is a 1904 reprint. I have never used human urine to make leather stain (page 46) but a must for the book shelf!

Accoutrements of the Riding Horse by Cecil G. Drew Pub. Seeley Services
My copy is beautifully printed on parchment-like paper, and, although not a technical manual, it is my favourite saddlery book to browse through.

The Heavy Horse, Its Harness and Decoration by Terry Keegan Pub. Pelham Books
An authorative book by a true heavy-horse man.

Encyclopaedia of Driving by Sallie Walrond Pub. J. A. Allen
If you want to know anything about driving it is here. A great reference book.

Making a Saddle Pub. Cosira (Council for Small Industries in Rural Areas)
To Handmake a Saddle by J. H. L. Shields Pub. J. A. Allen
Both these books deal, as their titles suggest, with the making of new saddles as a one-off production. Highly technical and correct, and a good read for anyone not familiar with the inside of a saddle.

Discovering Harness and Saddlery by Major G. Tylden Pub. Shires
A handy little paperback tracing the history of horse equipment through the ages in both Britain and abroad.

Saddler by Sidney A. Davis Pub. Shires
A fascinating account of the memories of a retired saddler. In it he documents the changes he has seen within the trade from the early 1920s to the present day.

GLOSSARY

Bearers These are the long, heavily padded parts of a saddle panel that distribute the riders weight each side of the horse's spine.

Broken top This describes a neck collar that opens at the top by way of a buckle and point to allow for ease of fitting over a horse's head.

Buckles, single The most common type is the bridle buckle which has only one metal bar to hold the leather in place. A loop or keeper is usually stitched next to the buckle to hold the leather point.

Buckles, double With this type of buckle, there are two metal bars to hold the leather in place. They are most usually seen on brass-mounted headcollars.

Fore piece The narrow piece of leather, about eight to ten inches long, that fits underneath the arch of the saddle and runs from one flap to the other. This might be omitted on a few low quality saddles.

Hames These metal, or wood and metal, fittings which run around the neck collar take the draft strain of a driving vehicle or heavy load, as when chain harrowing or pulling logs.

Keepers These are also called loops and are the bands of leather that hold a strap of leather in place once it has passed through a buckle.

Kip A cheap, low quality leather, only to be used where it is not going to come under any stress. This thin hide might also be referred to as basil.

Lacing This is usually carried out with a curved needle. It is a method of stitching used when there is access to only one side of a piece of work.

Lined and swelled A description of a good quality piece of driving harness strapping. With this method of manufacture, the strap has been made up from two thicknesses of leather with a third, very narrow, piece stitched in between them so producing a swelled effect. This is also a method of making nosebands for show bridles.

Mounts The colour and type of material with which the metal fittings are made. A good leather headcollar would be brass mounted, i.e. all metal fittings made from solid brass.

Piping A piece of round cord is covered by leather to form an edging along the outside of the panel on a saddle or driving pad.

Points, tree The ends of the saddle's wooden tree which fit into the padded panel.

Points, leather The part of a leather strap which has a series of holes to allow for adjustment when buckling up.

Pop stitching A method of hand stitching using one needle in which the stitch is formed on alternate sides of the work. This method is often used to hold piping in place or when relining a saddle or pad.

Pull up After 'taking down' or 'dropping' the panel of a saddle, you need to 'pull up' or restitch the two parts together again.

Returns The piece of leather approximately two inches long that folds over a buckle and is stitched together. With a single buckle, a keeper or loop is usually stitched into the return.

Roko These tongueless buckles are favoured for nylon headcollars, rugs and webbing driving harness, because they grip the webbing without piercing it.

Rugging A woven, blanket-type material, it is useful for relining saddles, driving pads and especially farm working collars.

Saddle seat This is the thin, yet strong, piece of leather that is stitched to the skirts of a saddle.

Shaft furniture On shafts used for private driving you will have shaft tips, tug stops and breeching Ds. These will be either brass or white-metal nickel, depending on the type of vehicle and harness being used.

Skive To thin down the thickness of a piece of leather with a knife, usually on a buckle return. This helps to keep the work neat and avoid unnecessary points of friction.

Splice A way of repairing a length of leather by skiving the ends of the old and new pieces of leather, laying one piece of leather on top of the other and stitching. For safety reasons, always allow a good length of overlap to stitch.

Take down Also known as to 'drop'. It describes the separation of the panel from the rest of the saddle.

Tugs, 1 The strong hoops of leather on a set of harness which hold the driving vehicle's shafts in place.
2 This term can also be used to describe the short buckle parts that form part of a breeching, neck collar or breastcollar.

Wale or forewale This is the circular piece of leather that forms part of a driving collar. It is the first part to be made and determines the finished size. Between the wale and the body of the collar, a pair of hames fits snugly into place.

Welting A very thin and narrow piece of leather that is stitched between the saddle seat and the skirt.

INDEX